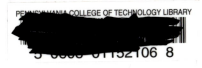

Aeromobilities

Aeromobilities is a collection of essays that tackle in many different ways the growing importance of aviation and air travel in the hypermobile, globalized world. Providing a multidisciplinary focus on issues ranging from global airports to the production of airspace, from airline work to helicopters, and from movement in airports to software systems, *Aeromobilities* seeks to enhance our understanding of space, time and mobility in the age of mass air travel. From São Paulo to Sydney, *Aeromobilities* draws on local experiences of airspaces to generate theory and research that are global in scope. It is the first book of its kind, bringing together a wide range of theoretical and methodological approaches to aviation and air travel in the social sciences and humanities, while emphasizing the central role of aeromobilities in contemporary social relations. In a world where virtually every aspect of social life is touched upon, in one way or another, by the complex global network of airline flows, with its large passenger aircraft and iconic international airports, *Aeromobilities* provides innovative analyses of some of the most fundamental and influential mobility networks of our time.

Dr Saulo Cwerner studied social sciences at the University of São Paulo, Brazil, international relations at the University of Leeds and cultural studies at Lancaster University before obtaining a PhD in sociology at Lancaster University. He worked at the Centre for Urban and Community Research, Goldsmiths College, and Lancaster University before joining the Centre for Mobilities Research at Lancaster University.

Dr Cwerner researched and published several articles in the fields of migration and refugee studies before conducting research on mobility, air travel and urban life. He is now developing further research on aviation and society from the perspective of mobility studies, with an emphasis on the environment and the politics of aviation.

Dr Sven Kesselring is a member of the TUM Transportation Centre in Munich, Germany, and the director of research of the mobil. TUM project, which is part of the Institute for Advanced Studies of the Technische Universität München. In 2008 he was a visiting professor for sociological theory at the University of Kassel.

He holds a PhD from the Ludwig Maximilians Universität München. Recently he received a grant from the Erich-Becker-Stiftung (Frankfurt Airport Foundation) for his work on airports and globalization.

John Urry is Distinguished Professor of Sociology at Lancaster University, UK, where he is Director of the Centre for Mobilities Research. His recent books include *Sociology Beyond Societies* (Routledge, 2000), *Global Complexity* (2003), *Tourism Mobilities* (edited with Mimi Sheller, 2004), *Mobile Technologies of the City* (edited with Mimi Sheller, 2006) and *Mobilities* (2007). He edits the *International Library of Sociology* and the journal *Mobilities*.

International Library of Sociology
Founded by Karl Mannheim
Editor: John Urry
Lancaster University

Recent publications in this series include:

Risk and Technological Culture
Towards a sociology of virulence
Joost Van Loon

Reconnecting Culture, Technology and Nature
Mike Michael

Advertising Myths
The strange half lives of images and commodities
Anne M. Cronin

Adorno on Popular Culture
Robert R. Witkin

Consuming the Caribbean
From Arwaks to Zombies
Mimi Sheller

Between Sex and Power
Family in the world, 1900–2000
Goran Therborn

States of Knowledge
The co-production of social science and social order
Sheila Jasanoff

After Method
Mess in social science research
John Law

Brands
Logos of the global economy
Celia Lury

The Culture of Exception
Sociology facing the camp
Bülent Diken and Carsten Bagge Laustsen

Visual Worlds
John Hall, Blake Stimson and Lisa Tamiris Becker

Time, Innovation and Mobilities
Travel in technological cultures
Peter Frank Peters

Complexity and Social Movements
Multitudes Acting at the Edge of Chaos
Ian Welsh and Graeme Chesters

Qualitative Complexity
Ecology, cognitive processes and the re-emergence of structures in post-humanist social theory
Chris Jenks and John Smith

Theories of the Information Society, 3rd Edition
Frank Webster

Crime and Punishment in Contemporary Culture
Claire Grant

Mediating Nature
Nils Lindahl Elliot

Haunting the Knowledge Economy
Jane Kenway, Elizabeth Bullen, Johannah Fahey and Simon Robb

Global Nomads
Techno and New Age as Transnational Countercultures in Ibiza and Goa
Anthony D'Andrea

The Cinematic Tourist
Explorations in Globalization, Culture and Resistance
Rodanthi Tzanelli

Non-Representational Theory
Space, Politics, Affect
Nigel Thrift

Urban Fears and Global Terrors
Citizenship, Multicultures and Belongings After 7/7
Victor J. Seidler

Sociology through the Projector
Bülent Diken and Carsten Bagge Laustsen

Multicultural Horizons
Diversity and the Limits of the Civil Nation
Anne-Marie Fortier

Sound Moves
IPod Culture and Urban Experience
Michael Bull

Jean Baudrillard
Fatal Theories
David B. Clarke, Marcus A. Doel, William Merrin and Richard G. Smith

Aeromobilities
Ed by Saulo Cwerner, Sven Kesselring and John Urry

Aeromobilities

Edited by
Saulo Cwerner, Sven Kesselring
and John Urry

Routledge
Taylor & Francis Group

LONDON AND NEW YORK

First published 2009
by Routledge
2 Park Square, Milton Park, Abingdon, Oxon OX14 4RN

Simultaneously published in the USA and Canada
by Routledge
270 Madison Avenue, New York, NY 10016

Routledge is an imprint of the Taylor & Francis Group, an informa business

© 2009 Saulo Cwerner, Sven Kesselring and John Urry

Typeset in Times New Roman PS by
Florence Production Ltd, Stoodleigh, Devon
Printed and bound in Great Britain by
MPG Books Ltd, Bodmin, Cornwall

British Library Cataloguing in Publication Data
A catalogue record for this book is available from the British Library

Library of Congress Cataloging in Publication Data
Aeromobilities/edited by Saulo Cwerner, Sven Kesselring and John Urry.
    p. cm.
    Includes bibliographical references.
    1. Aeronautics – Social aspects.   2. Air travel – Social aspects.
    I. Cwerner, Saulo, 1963–.   II. Kesselring, Sven, 1966–.   III. Urry, John.
    TL553.A348 2009
    387.7 – dc22   2008048253

ISBN10: 0–415–44956–1 (hbk)
ISBN10: 0–203–93056–8 (ebk)

ISBN13: 978–0–415–44956–4 (hbk)
ISBN13: 978–0–203–93056–4 (ebk)

Contents

Contributors

Peter Adey is a Lecturer at the School of Physical and Geographical Sciences at Keele University, United Kingdom, where he teaches human geography. His research interests include security, software, surveillance, power and architecture, and he has published extensively on airports, aviation and air travel.

Lucy Budd is a Researcher in the Geography Department and the Globalization and World Cities (GaWC) Research Network, Loughborough University, United Kingdom. The focus of her research is on the multiple geographies of British airspace, its production, representation and use.

Saulo Cwerner is a Researcher at the Centre for Mobilities Research, Lancaster University, United Kingdom. His main research interests include aviation, mobilities, globalization and migration, with a particular focus on the politics of air travel and the cultural representation of flight.

Ben Derudder is Professor of Human Geography at the Geography Department, Ghent University, Belgium. He also heads the Transport and Production section of the GaWC Research Network, where he is a Research Fellow. His research interests include conceptualizations of transnational urban networks.

Martin Dodge is a Lecturer in Human Geography at the School of Environment and Development, University of Manchester, United Kingdom. His major research interests include geographies of cyberspace, surveillance and securitization. He is also co-author of the *Atlas of Cyberspace*.

Guillaume Faburel is Associate Professor at the Institut d'Urbanisme de Paris and Researcher at the Centre for Research on Planning: Land Use, Transport, Environment and Local Government, University of Paris 12, France. He has researched and published extensively on aviation and air travel, particularly on sustainable mobility, aviation policy and governance, and the local impact of aviation noise.

Gillian Fuller is Senior Lecturer in Media at the School of Media, Film and Theatre, University of New South Wales, Australia. She has published extensively on the social semiotics of movement and order, and is co-author of *Aviopolis: A Book about Airports*. Her current research interests include emergent mobile architectures.

Ross Rudesch Harley is Associate Professor and Head of the School of Media Arts at the College of Fine Arts, University of New South Wales, Australia. He is also an artist, writer and educator in the field of new media and popular culture, with research interests and expertise in digital video, media arts and network theory, among other subjects. He is also co-author of *Aviopolis: A Book about Airports*.

Sven Kesselring is Senior Researcher at the Institute for Transportation, and Director of Research of the mobil.TUM project at the Institute of Advanced Studies, Technische Universität München, Germany. He also co-ordinates the Cosmobilities Network and has several works published in the fields of mobility, risk and globalization.

Rob Kitchin is Professor of Human Geography and Director of the National Institute for Regional and Spatial Analysis at the National University of Ireland, Maynooth. His research interests include the geographies of cyberspace, space and place, and he is co-author of the *Atlas of Cyberspace*.

Claus Lassen is Assistant Professor at Aalborg University, Denmark. His research interests include mobility, air travel, work, cities and the knowledge industries.

Lisa Levy is Researcher at the Institut d'Urbanisme de Paris, University of Paris 12, France. Her work focuses on aviation, sustainable development and territorial conflicts around airports.

Peter Peters is Senior Lecturer at the Faculty of Arts and Culture of Maastricht University. He has published on music, time, and travel in technological cultures. His research interests include the mobilities turn in social sciences, border studies, and globalizing art worlds.

John Urry is Distinguished Professor at the Department of Sociology and Director of the Centre for Mobilities Research, Lancaster University, United Kingdom. He has published extensively in the fields of social theory, social change, urban and regional research, travel and tourism, and the environment.

Nathalie Van Nuffel has worked as Researcher at the Department of Geography, Ghent University, Belgium. Her main interests include spatial planning, housing and commuting.

Frank Witlox is Professor of Economic Geography at the Geography Department, Ghent University. His main research interests include transport economics and geography, world cities and globalization, and urban planning. He also teaches and researches at the Universities of Antwerp and Hasselt, and is a research member of the GaWC Research Network, Loughborough University.

Preface

The idea for this book originated during a conference organized by the Centre for Mobilities Research (CeMoRe) (Lancaster University) and the Cosmobilities Network, held at Lancaster University in September 2006. *Air-Time Spaces: Thinking through Mobilities Research* brought together a multidisciplinary set of academics from several European countries as well as Australia to discuss the place of aviation and air travel in contemporary social sciences and humanities. Around forty scholars gathered to discuss a dozen or so papers by people who have been conducting new research on these subjects, some of whom had published pivotal work on airspaces and air travel.

Despite the wide-ranging topics and disciplinary affiliations, it became clear to the organizers of the conference that virtually all participants had something fundamental in common: a belief in the importance of studying aeromobilities for understanding key issues of our time, from globalization to the environment, from security to work, from cities to personal relationships. But it also became clear that for many the lively exchange of ideas was revealing new and previously unsuspected perspectives on air travel and aviation. As we learned from each other, we realized the need to pull our efforts together in order to boost the development of aeromobilities research. The conference organizers quickly identified that, outside the discipline of aviation history, there was no single published volume that brought together a range of different perspectives and approaches to aviation and air travel in the social sciences and humanities. The idea for this book was born.

Any such volume is bound to leave out much more than it includes. For instance, the issues of security and the environment figure prominently in current work on aviation in the social sciences and related disciplines. Because of that, these topics received less systematic attention in this book. Yet the politics of air travel, the production of airspaces, work and social relations, among other subjects, get diverse treatment here. We hope we have succeeded in bringing together an array of fundamental works on aviation and air travel. This collection of essays will hopefully pave the way for more innovative research to 'take off' in this area.

Much of the research and theory showcased in this volume has been a more or less direct outcome of the recent 'mobility turn' in the social sciences, of which both the CeMoRe (Lancaster University) and the Cosmobilities Network have been major exponents and breeding grounds. Apart from its impact on research, theory and policy in the fields of mobility, travel and transportation, mobilities research has provided an evolving and multifaceted paradigm for social scientists, especially those working around the tenuous boundaries that still exist between the social science disciplines and their subject matter. This book implies in strong ways that the study of *aeromobilities* can have a similar impact. Aviation, air travel and all related facts have become such a large, central and all-encompassing feature of contemporary social life that they can no longer be taken for granted. There is a growing need for unpacking their role in shaping social relations, and a concomitant need to bring together and articulate the largely isolated data and analyses that are coming out of individual research. In this way, and despite comprising separate and independent essays by contributors from different academic disciplines, this book carries a soft programmatic message. At a time when the aviation industries, governments and civil society are awakening to current and future impacts of air travel, academics, researchers and students need to pay attention not only to those machines flying over our heads, but also to the massive infrastructures (moorings) that they necessitate, and the social relations that they enable and permeate, on the ground.

Finally, the title of this book, *Aeromobilities,* also draws inspiration from the mobility turn in the social sciences and the humanities. In fact, the study of aviation and air travel lies at the core of the development of mobilities research in general. It also intimates a plurality of mobilities, from passenger jets to space stations, from aerial photography to paragliding, from rooftop helipads to aerial bombardments. We can no longer ignore what makes these mobilities possible (including plentiful cheap fuel), nor their impact on communities, territories and social groups globally. Bearing on a diversity of perspectives, both theoretical and methodological, and suggesting a number of innovative representational strategies, *Aeromobilities* calls for a collective effort to expand our knowledge of airspaces and the social relations they enhance and make possible. Certainly, different perspectives are opened by this book, but one of its major undercurrents is the understanding that our relationship with aeromobilities is deeply *political*. And, without knowing what lies above us, we have very little scope for bringing it under democratic control.

Acknowledgements

We would like to thank everyone who was involved with the *Air-Time Spaces* conference. Most of the contributors to this volume were present there, but others also made it possible through their diligence and creativity. Pennie Drinkall, Laura Watts, Yoke-Sum Wong, Ann-Marie Fortier and everyone at CeMoRe made invaluable contributions to the organization of the conference, as did Gerlinde Vogl, Sanneke Kloppenburg and everyone at the Cosmobilities Network. We would also like to thank David Pascoe, Dawn Lyon and all the other participants in the conference for their thoughtful presentations and discussions, which also inspired us to proceed with this project. Without this collective effort, the foundations for this book would not have been laid. We would also like to thank our editors at Routledge, Ann Carter and Miranda Thirkettle, for their crucial help and unending patience and understanding.

Last, we thank all readers of this book for their interest and would like to encourage their comments and feedback, which could help us to improve and propel aeromobilities research into a substantial strand within social science-based mobility research in the future.

Introducing aeromobilities

Saulo Cwerner

> It is not worth while to conquer space if we cannot devour it.
> (Jules Verne, *Robur the Conqueror*, 1886)

In 1946, the eminent American sociologist William Fielding Ogburn published *The Social Effects of Aviation*, an early systematic attempt at technological and social forecasting, looking at the ways through which aviation would impact on social change in a number of areas, from leisure and family life to population, cities and international relations. Ogburn's specific predictions are not an issue here, and they have been proven flawed in many respects (Richter 1991). For instance, based on contemporary predictions about the spread of automobility, Ogburn confidently stated that aviation would lead to a drop in the rate of population growth, because families would put the ownership and operation of personal aircraft ahead of procreation. In part, this early vision of generalized aeromobility was clearly linked with efforts towards producing a 'poor man's aeroplane', spearheaded in the 1930s by Eugene Vidal at the helm of the Bureau of Air Commerce in the USA. The $700 aircraft would cost the same as a nice car and would put aeromobility within the reach of hundreds of thousands of American families (Moore 2006).

Nevertheless, Ogburn's study, although tainted by dated perspectives on technology, innovation and society, and based on research undertaken mostly before World War II (by the time the study was published in 1946, the Douglas DC-6 was about to supersede de DC-3 in mass air transportation), was important in that, for the first time, the relations between aviation and society were probed from the perspective of social research. This was a breath of fresh air within a field where inventors, industry tycoons, promoters and other aviation pioneers held sway. For several decades after the Wright brothers' first successful endeavours in the very beginning of the twentieth century, aviation *was* the future, and aeroplanes were seen as 'prophetic machines, promising enhanced mobility, enlarged prosperity, cultural uplift, and even social harmony and perpetual peace in an emerging "air age"', suffused by the other-worldly excitement of a new secular religion, the 'winged gospel' (Corn 2003: x). As a result, before the age of mass air travel, there was little to commend the study of the concrete ways through which aeromobilities helped shape modern social life, which makes Ogburn's effort look decidedly ahead of its time.

To be sure, both the cultural history of early flight and the thrust of whole nations into air-mindedness in the first half of the twentieth century have now been well documented and researched (see, for instance, Wohl (2005), Fritzsche (1992)). However, while social and cultural historians have provided both new perspectives and comprehensive narratives on aviation and air travel (Bilstein 2001), fifty years after Ogburn's path-breaking book, aviation and air travel still receive scant attention in the social sciences. Nevertheless, the last decade has witnessed a surging and unprecedented interest in these subjects, in part as a result of the 'mobility turn' and a new mobility paradigm that have yielded a more systematic interest in physical travel and its relationship with other mobilities, networks and systems (see Urry (2000; 2007); Sheller and Urry (2006); Larsen *et al* (2006) and Cresswell (2006)). The aim of this book is to showcase recent work on aviation and air travel in the social sciences and make a case for the establishment of aeromobilities as both a distinctive field of research and a key issue in the future development of social theory in an increasingly complex, mobile world.

Before introducing the various chapters that compose the book, this introduction will seek to answer two fundamental questions: (a) why aeromobilities matter to social scientists and (b) how aeromobilities research can contribute to contemporary social theory and knowledge in the social sciences. It is the editors' belief that aviation and air travel are not simply appendages to a world of globalized and hypermobile social relations, networks and risks. Not only are they *central* to this world but also a key component of both the material processes and debates that are shaping the future of society and the environment.

Why do aeromobilities matter?

The study of mobilities addresses the large-scale, long-distance physical movement of people, goods and ideas. Much of this movement nowadays occurs by air, using more often than not large passenger aircraft. A global total of 2.75 billion passengers is forecasted for 2011, following an expected average yearly increase of 5.1 per cent from 2007. Although this rate is lower than the average for the previous five-year period, the global airline industry will be boosted by increased levels of participation from China, India, Russia and the Middle East.[1] The global numbers obviously hide large discrepancies and inequalities between countries and regions. While about half of the population of the United Kingdom are believed to fly at least once a year (Department for Transport 2003: 21), a recent survey in Brazil suggested that only 8 per cent of the population travel by air, even only occasionally, and this in a country where long distances have created a reasonably well-developed domestic air transportation system.[2] Even more striking, in India only one in forty-four people travelled by air in 2006 (Airbus 2006: 9). It is a simple truth that, as long as current environmental, safety and security concerns are managed satisfactorily, the aviation industry has a massive growth potential, especially in developing countries, where emerging middle-classes should provide steadily growing demand for both domestic and long-haul air travel. Industry assessments are clear: 'Lower air fares, more service, and the increasing value of time in these emerging countries, are inexorably pulling traffic from buses and trains into aircraft' (Airbus 2006: 9–10).

The numbers are superlative, be they of passengers, aircraft, freight or airmail, not to mention the large airports that have come to define the cities and regions that they serve, even if their principal role is that of a transfer hub. This is undoubtedly the core of contemporary aeromobilities, but to them we can also add the less well-known field of general aviation, with its executive jets, leisure aeroplanes, helicopters, hot-air balloons and a myriad of other machines and social activities that impart a distinctive three-dimensional character to work, recreation and the environment. And, we must not forget, military aeromobilities that have increasingly influenced military strategy and international relations, especially with the recent debate over precision air weapons (Pape 2004, McPeak and Pape 2004) and whether wars can be won solely through air power (Byman and Waxman 2000), not to mention the political and ethical fallouts of aerial bombardments (Grosscup 2006). The airspaces of military aeromobilities affect countless communities in peace as well as war, both within nations and in foreign lands, often drawing civil aviation itself into military, strategic and diplomatic disputes (Dobson 1991).

What do we make then of this diverse, evolving aggregate of mass air travel, complex international airports, on-demand aviation, supersonic bomber jets, news helicopters, nostalgic air shows and space travel, which, among many others, constitute the multiple world of aeromobilities? Why do we need to pay attention to that which moves in the air? In his chapter in this volume, John Urry invites us to think, counterfactually, of a world without these aeromobilities. Is it at all possible to even imagine, from the privileged viewpoint of the twenty-first century, family life, cities, work, popular culture, war, migration, education, leisure, tourism, communication and government (the list could go on) in a world without aviation? Possibly not, so intertwined with air travel modern life and globalization have become. So what is so specific and irreplaceable about aeromobility that has made air travel constitutive of modern social life?

Answering this question poses two problems. One must resist any form of technological determinism that would look at the development of aviation as a tale of progress, to which society and its institutions strive to adapt. And one must also avoid romanticizing aviation to the point of overstating its impact upon modern life. The new mobilities paradigm allows us to look at aeromobilities in their relations with various social networks and systems, therefore *grounding* or *embedding* them in processes whereby these mobilities, and their own distinctive spaces, networks, systems and environments, are effectively produced, reproduced, performed and regulated. Technologically speaking, aeromobilities are part and parcel of modern life in the same way as automobility, surface public transport or sea travel. One only needs to remember that barely a generation elapsed between the invention of the modern car and the Wright brothers' early successes with flying machines. And the same year, 1914, marked both the symbolic birth of the mass production of cars with the adoption of Henry Ford's revolutionary model of car production *and* the first ever commercial flight. As we acknowledge the rise of a culture of air travel (Gottdiener 2001), whereby more of us are literally and routinely living greater proportions of our lives in airspaces (between airports, aircraft, flight paths and the social worlds that they connect), the analysis of aeromobilities must account for the complex interdependencies between different mobilities, networks, systems, institutions, risks, cultures and territories.

On the other hand, it is also necessary to acknowledge the specificities of aeromobilities. And what is so distinctive about them is *altitude* and *speed*. Altitude allows a novel perspective on land and the environment, one that has been used successfully in a variety of fields. Having produced some of the most iconic representations of the natural and human environments (not to mention the Earth herself), the *aerial view* comprises a set of spatial and representational practices that have changed the dynamics of contemporary war, mapping, surveillance, art, journalism, urban planning and government, to name a few (Boldrick 2007). But the altitude made possible by aeromobilities is not simply about cultural representations, whatever their profound impact on social life may be. Aeromobilities also provide aerial access to places that would be virtually unreachable by other means, proving a key element in rescue and other emergency operations, exploration and travel. With helicopters at the forefront, aeromobilities have recast the significance of natural and other disasters, and reconfigured the balance of risks in contemporary environments. Both the *aerial view* and *aerial access*[3] have promoted a globalization that includes remote areas alongside the major nodes in the global economy and provided new, three-dimensional features of contemporary social life.

More important, and by no means unrelated, is *speed*. As a function of time and distance (providing simultaneously time savings and the conquest of longer distances), aeromobilities have been a key agent in the shrinking of the world, enabling social relations and networks to be maintained over longer distances and at greater regularity and pace. As it has been shown with regards to military aircraft (Law 2002: 31), speed is not simply a technical problem, but engenders particular performances of 'heroic agency'. This 'heroism', which predicated the early decades of powered flight and still permeates discourses of military aircraft, is still distinctively associated with the long-haul holiday with its discourses of self and discovery (Davidson 2006: 31–2), which reproduces in part the heroic nature of early flight in the context of high-speed passenger jets.

Aeromobilities have helped social networks and ties to be formed over long distances. In particular, advances in air travel and modern communications have boosted transnational ties between immigrants and their countries of origin. This is not new, of course, as travel and exchange of letters character-ized earlier immigration as well: 'what is new is the frequency and greater ease with which such back-and-forth contact is maintained, resulting in trans-national families [and] global communities' (Ebaugh and Chafetz 2000:143). Aeromobilities also have changed the shape of contemporary work (Lassen 2006), providing new opportunities for business, especially over long distances. Thus aeromobilities cannot be dissociated from the places, people and organizations they connect, the distances that they bridge and the speed with which they distribute people, objects and information across these distances. Survey data from 2006 have shown that, in the United Kingdom, 90 per cent of air travellers flew for leisure purposes (for instance, to go on a holiday), while 15 per cent did so to visit friends and family, and 13 per cent made business trips.[4] Globally, there is a great diversity in the purpose of air travel for individuals, reflecting national specificities such as geography and income levels. Thus, while in Malaysia and Thailand over 20 per cent of air passengers travel on business, the figures are 6 per cent and 7 per cent, respectively, for the Germans and French.[5] Families, communities, workplaces and organiza-tions are extended in space, as more places and people are visited and consumed, more frequently and in decreasing times.

It is almost meaningless to analyze the dynamics and trends of work, business, family and personal relationships, higher education, professional sport and recreation, popular culture, tourism, diplomacy, and virtually all significant areas of contemporary social life, without taking into account the particular and distinctive time/spaces created by aeromobilities. The dispersal of organizations, networks and systems across space has drawn from, and then fed back into, evolving systems of mass air travel, which eventually led to the production of specific cognitive forms of space/time compression (Gottdiener 2001: 141), which could be termed an aeromobility *habitus*. This is not only a feature of frequent flying, but can also be seen in the less frequent holiday by air. The growth of low-cost airlines has exacerbated the *will to fly*, bringing new cities and regions into the global systems of aeromobility, with leisure passengers now forecasted to take an increasing number of multiple short holidays (Mason and Alamdari 2007). According to the global survey conducted by Synovate in 2005, 75 per cent of respondents agreed with the statement that an 'airline trip is just like a bus trip' and said that they wanted to 'get there fast and cheap'.[6]

It is this *habitus*, the readiness to fly, increasingly boosted by various net-works and systems, such as airports, scheduled aviation, global corporations and a myriad of tourist destinations, that makes the study of aeromobilities even more relevant and interesting. Never mind the tales of the golden age of air travel, where things in the cabin 'were most luxurious because only the wealthy and VIPs could afford air travel' with the 'Silver Wing' service provided by Imperial Airways in the 1930s (Quinn 2003: 10; see also Hudson 1972: 33–61). Aeromobilities are now even more important because they have become routine, matter-of-fact, effectively *banal*. And this is the point where globalization literally takes off, where both the compression and the distantiation of time and space can be not simply conceptualized as long-term historical processes, but *observed* as everyday performances by an increasing array of social actors, networks and objects.

In Nazi Germany, the whole German population was systematically thrust into air-mindedness. As the 'aeroplane expressed the will to power over the material world', registering power and possibility, aeromobility became a political machine (Fritzsche 1992: 185). At the beginning of the twenty-first century, the fascination with air travel (and its ambivalent effects) has perhaps diminished, with economics taking over. Instead of places of spectacle, imagination and mobilization, airports became the ' "concrete" materialization of "transit" and "neutrality" ' (Pascoe 2001: 229). To understate this shift would be theoretically dangerous: aeromobilities now permeate social worlds and everyday lives more than ever before, and this is why they matter, not just to the specialist or enthusiast, but also to the social scientist going about his or her business of understanding the various facets of contemporary social life. And because of that, aeromobilities must not be seen as a mere manifestation of something else. The constant, rapid, ephemeral and, at the same time, pervasive count of air departures and arrivals in the contemporary world are effecting a massive reconfiguration of space and territory, even if it is not much more than a radical acceleration of what was already happening on roads, tracks and water. But this radical acceleration, this hourly disturbance of community, fixity and familiarity, and their transposition to different (often far-away) places, needs to be unpacked and rigorously analyzed in its multiplicity, complexity, ambivalence and contradictions. Without systematic analyses of aeromobilities, scholars in the humanities and social sciences may miss out on much-needed insights about the shape of things to come.

Researching aeromobilities

Now, even as the case for aeromobilities research is successfully made, how do we proceed to study them, so that aeromobilities research can have a lasting impact in the social sciences and social theory? The answer to this question resides in the growing body of work on aeromobilities that has been developing in (and in-between) various disciplines, of which this book presents some prime examples.

One of the most important issues regarding aviation is its sheer size. In 2005 the industry itself claimed that its global economic impact amounted to close to US$3 trillion, equivalent to 8 per cent of the world's GDP, employing around five million people directly in its airline, airport and aerospace sectors, without counting the several million purported indirect jobs, many of them created by the growing role of air travel in the tourist industry

(ATAG 2005: 2), (see also York Aviation (2004)), which makes tourism a central issue in aviation policy worldwide (Forsyth 2006). There is great controversy over the real economic impact of the aviation industry, however, with several academics and environmental groups questioning industry and governmental data and analyses, particularly in debates over issues such as job creation and the balance of regional spending and investment resulting from air travel expansion (Whitelegg 2003, Boon and Wit 2005).

Despite such controversies, it would not come as a surprise that the vast majority of the literature on aviation comes from the fields of economics and management, including the impact of regulation and deregulation. Although these areas are not the focus of this book, there is no doubt that social scientists at large would benefit immensely from understanding the economic and organizational strategies of the various sectors of the aviation industry. Key to the expansion of air travel in the past thirty years has been the changing nature of the regulatory framework for air travel, in particular the deregulation of the industry, which began in the late 1970s in the USA, leading to an increasingly global competition between largely privatized airlines, airports and aircraft manufacturers. The reorganization of the aviation industry that followed, which included the rapid growth of low-cost carriers, the advent of transnational alliances between major airlines, and the rise of the regional jet, has had a great impact on patterns of aeromobility: the historical drive towards a virtual 'open skies' regulatory framework has boosted the *will to fly* and led to a radicalization of time/space compression and the obliteration of more territorial boundaries, with new regions and localities brought into systems of aeromobility (Bennett 2006: 31ff). Deregulation, liberalization and the new aviation regime have been particularly crucial for the tourism industry world-wide (see Cornelissen 2006: chapter 6).

The rise of the aviation industry has also seen it figure prominently in recent environmental and political debates. On one side of the debate, the commercial aviation industry, represented by international organizations such as the International Air Transport Association (IATA – the main global airline representative) and Airports Council International (ACI – which represents airport operators), puts forward an environmental agenda based on technological and operational improvements, while alliances between local groups and global environmental groups favour more restrictive policies on gas emissions and aircraft noise in their vision of sustainable development for an industry that is believed by some to become responsible for 15 per cent of climate change by 2050 (see Upham *et al.* 2003; Whitelegg and Cambridge 2004). This has led to some research particularly on the politics of airport noise and airport expansion, the result of which has been an incipient debate about new forms of airspace governance, whereby commercial and environmental concerns clash in what remains one of the key political and organizational areas where there is a considerable democratic and participatory deficit (Griggs and Howarth 2006).

Crawling along the established studies of aviation economics and management, and the fast growing field of aviation and the environment, social research into aviation from other disciplines in the social sciences has only recently made serious attempts to catch up. Because of their strong visibility as icons of the global order, and perhaps also as a result of the practical difficulties in carrying out research on mobilities in general, a great deal of this recent work has focused on airports. Cultural historians have provided thorough analyses of the development of airports, in terms of both their architectures and their sites in popular culture and imagination (Pascoe 2001, Gordon 2004). As the site of global hypermobility *par excellence*, the airport has also invited cultural analyses looking at how information and architecture combine to produce particular logistics of mobility (Fuller and Harley 2005, Fuller 2002) and spaces of consumption (Lloyd 2003). This focus on airports as a distinctively modern urban form in their own right has complemented urban economic analyses that reveal the condition of airports as poles of strategic regional development (COFAR 2001), although it is important to distinguish between different types of airport, depending on size and on the amount of transfer passengers they host, when locating aeromobilities within regional development. The fact is that airports have long moved from the fringes of cultural analysis into the centre of planning discourse (Kasarda 2006), with the 'aerotropolis' providing a new utopia of urban development around airports.

However, beyond sites of global hypermobility, managed transit and urban/regional development, airports and aircraft have also developed into dynamic instruments of surveillance, security, control and risk in a long historical process that long predated, but culminated symbolically in, the 2001 terrorist attacks against the World Trade Center in New York City (see Wilkinson and Jenkins (1999)). The new regime of surveillance and security in airports is now being well researched and documented (Lyon 2006, Adey 2004). And in parallel with studies of the politics of aviation risk and security, the wholly self-contained discipline of 'human factors' applied to aviation has developed, providing a good deal of insight into human behaviour in aviation environments, human-machine interaction and management systems, which has had great impact on design, technology and training (Harris and Muir 2005). Despite data showing that air travel is among the safest forms of travel, high-profile accidents have always conjured up images of risk and danger that accompany aviation to this very day of mass air travel. It is not surprising that among topics of interest in the study of both crew and passenger behaviour, air rage, the impact of aviation disasters and hostage behaviour figure prominently (Bor 2003; Bennett 2001, 2006).

Research into aviation and air travel has been quite extensive. However, it has developed very much along tight disciplinary lines, resulting in a dearth of attempts at integration of these multiple empirical and theoretical fields, however tentative they may be (Gottdiener (2001) is a notable exception). Moreover, some of the major social sciences, such as sociology, geography, politics and anthropology, have not yet integrated the study of aeromobilities into their main theoretical and empirical foundations. But, with a developing number of scholars demonstrating an interest in the subject, resulting in an incipient dialogue with areas where air travel and aviation have been systematically researched (for instance, in centres of transport studies and research), time is surely ripe for setting out a comprehensive agenda for aeromobilities research that builds on current knowledge but also aims at both integrating and diversifying research topics, methods and strategies.

The contours of this research agenda will be designed by deepening the exchange of ideas in the fields of study outlined here, and by disseminating crucial aeromobilities research among the wider research and teaching community in the humanities and social sciences. However, based on current research and theoretical work, it is possible to identify a number of key strategies for the future study of aeromobilities. *First*, aeromobilities research needs to be transdisciplinary (or post-disciplinary). Social scientists will at their own peril ignore developments in their neighbouring departments. Although this may now hold true for any social research endeavour, perhaps no other field of enquiry requires more attention to the complexity of human life than aeromobilities research. In very few other fields would one encounter such intricate relationships between social life, technology and the environment, or more contradictory sets of mobilities *and* immobilities. The challenges of aeromobilities involve, *simultaneously and to a similar degree*, technology, community, governance, time/space perceptions, social interaction, urban development and the environment, among other issues. Isolating one or another through deference to traditional disciplinary foci and research agendas will diminish the impact of aeromobilities research. Post-disciplinary science is not an easy task, and does not come as a ready-to-use package. It will necessarily be the result of an evolving dialogue between different theoretical perspectives and traditions in the social sciences and beyond, to which this book aims to be a key contributor.

Second, aeromobilities research must be able to integrate different empirical fields and provide a multiple focus to research. As air travel enables family and personal relationships to extend over long distances, there is little point in debating the externalities of aviation, that is, its impact on local and global environments, without taking into account the increasing need to fly in order to accomplish basic social aims or perform key social roles. Similarly, the production of safety and security in airspaces, through diverse systems of air traffic control and passenger surveillance, can no longer be dissociated from the politics of surveillance and changing forms of governance and jurisdiction of airspaces. Because aeromobilities research demands the integration of topics studied separately, it will require new methodological wisdom, including innovative visual methods that can render hypermobility in ways that traditional narrative or statistical methods cannot.

Third, despite the continuing need to study the most celebrated spaces of aeromobilities, such as airports, aircraft and tourist destinations, research must also endeavour to visit and revisit everyday life. We need to understand a lot more the role that air travel plays in daily life, in the fashioning of personal relationships, and the construction and maintenance of social ties. While airports encapsulate much of what is characteristic of our complex, global and hypermobile age, there is an urgent need to learn more about how ordinary people imagine, dream about, plan and, more than just occasionally, discard, air travel, and how they react to commercial strategies of the aviation industry and new regulatory frameworks, etc. In short, what makes people travel by air, and what keeps them from flying (see Shaw and Thomas (2006)). Aeromobilities need to provide a much bigger empirical focus on the *will to fly* than we have at the moment, building on the work by cultural historians, but using social research methods to describe and analyze what I call the aeromobility *habitus*.

Fourth, aeromobilities research needs to link up with the wider field of mobilities studies in order to help provide a more comprehensive picture of our complex social world. On the one hand, the air transportation of people is just one aspect of aeromobilities, as described above. Aviation has revolutionized the mobility of information, goods, cultural artefacts, letters, performances and much more, and all these aeromobilities must be analyzed, often in conjunction with each other. On the other hand, aeromobilities are related in several different and distinctive ways to other forms of mobility, sometimes as rings in a chain of intermodal transportation systems, but often in a system of competing modes and scales of travel, including competing places and destinations.

And *fifth*, aeromobilities research must not forget the differentiation and segmentation of social life along economic, gender, ethnic, racial, age, physical ability and other lines. However tempting it is to see aeromobilities as representing general trends in social life, we must avoid generalizations that could significantly mask crucial differences through which systems of aeromobilities impact different social groups and/or categories. Similarly, we need to pay attention to the possible ways through which aeromobilities help produce certain identities and inequalities. Hochschild's classic study of the emotional labour of air hostesses is a good example of the gendered nature of air travel that has characterized it throughout its history (Hochschild 1983), but the history of aviation is also a history of both exclusion and liberation as far as the participation of women goes (Douglas 1990). We do not know much about the relationship between social class and air travel, and there is considerable need for research in this area. Survey data indicate, for instance, that the vast majority of low-cost flights are taken by more privileged social classes (see CAA (2004)), and that families with second homes tend to fly several times in a year, which suggests the role of low-cost air travel in deepening social inequalities rather than democratizing aeromobility. And the role of race, ethnicity and nationality in recent security and surveillance trends associated with air travel, particularly debates around the use of racial profiling, has provided further indication of the need to study the mutual implications of identity and aviation.

The social sciences as a whole need to take aeromobilities seriously, harnessing the work developed within social and cultural history, economics, management and the applied 'human factors' sciences, for the benefit of social research's ultimate agenda: the understanding of social life, its problems, limits and excesses. The historical vocation of social research, that is, its unmitigated concerns with issues such as power, inequality, community, individualization, values and communication, as well as its contemporary focus on the evolving complexity of technology, risk, globalization and the environment, requires that we look at aeromobilities as a central feature of contemporary social life that provides people with both escape routes and unprecedented control. It is perhaps this uncompromising ambivalence of contemporary aeromobilities that makes them such a fascinating topic for social research. The various contributors to this volume have taken to the task of addressing this deficit from a multitude of perspectives. By bringing these efforts together for the first time in a book, *Aeromobilities* lays down a new foundation stone in the study of global mobilities in the twenty-first century.

Contributions to the book

The first part of the book, 'Aeromobilities, globalization and social theory', provides strong foundations for the study of aeromobilities in the context of contemporary social theory and studies of globalization and risk. In 'Aeromobilities and the global', **John Urry** examines the centrality of air travel to the new global dis/order, looking in particular at the way that people and systems create new time-distantiated social networks through flying. Following a periodization of airspaces, Urry shows how the complexity and tightness of the various systems that enable air travel produce very distinctive environments of risk, and that it is the increasing sophistication of such systems that both allows relatively risk-free, regular air travel, while at the same time making it vulnerable to various disruptions. As a result, air travel has come to epitomize the global order, and Urry goes on to suggest a comprehensive theoretical and research agenda that takes account of the 'umbilical cord' between air travel and the global. These 'lines of flight' intimate the complexity of air travel, revealing, among other things, the role of air travel in synchronizing social networks and producing not only flows and mobilities, but also new modes of dwellingness, inequalities and power. However, Urry cautions, despite its potential for expansion and diversification, the very fact that 'air flights are central to performing the global order', and its role in forging the increasing interconnectedness of places and networks in this planet, subject air travel to the limitations of global climate change, of which it may still become one of the first distinctive casualties.

In 'Global transfer points: The making of airports in the mobile risk society', **Sven Kesselring** analyzes the contemporary international airport in order to bring to the fore a fundamental dimension of aeromobilities: the local environments and spaces that not only make global movements possible, but also redefine cities and regions around the world. By looking at airports as 'interfaces between the territorial and the global' and as 'paradigmatic localities', Kesselring is able to locate trends in the evolving global order of the 'glocal' politics that has developed around many airports in recent years. Most crucially, Kesselring sees airports as shifting gears between the local and the global. By grounding analyses of the global role of airports and their development into 'global transfer points' at the territorial level, Kesselring shows that the globalization made possible by aeromobilities does not evolve in a linear fashion, but is increasingly dependent on spatial and territorial conflicts. In this sense, airports are privileged sites of the dialectics of motion and fixity that characterize global society. Aeromobilities theory and research needs to take into account the new forms of politics and governance that are developing around airport expansion, and their relationship with the mobility imperatives of global networks. Using a range of empirical illustrations, Kesselring provides an essential theoretical account of the complex relationship between the spaces, politics and globalization of contemporary aeromobilities.

The second part of the book, 'The Production of Airspaces', deals with the different spatial dimensions of aeromobilities. The spaces of air travel are imagined, planned, designed, monitored, that is, *produced*, in a variety of ways, by a variety of people and organizations, and through a variety of methods. Moreover, these spaces (and times), from the queues before check-in desks, through the virtual world of airline software systems, to the air routes and corridors traversed by aircraft, are all related to each other in increasingly complex and sophisticated ways. This part showcases groundbreaking work that focuses on spaces of aeromobilities with which most of us remain unfamiliar.

In '> store > forward >: Architectures of a future tense', **Gillian Fuller** provides a theoretically informed examination of waiting in the context of the complex logistics of air travel. The flows of aeromobility are here conceptualized like information flows, and Fuller's overall aim is to analyze the forms through which movement in airports becomes informationalized and pre-emptively anticipated. Fuller begins her analysis by considering the airport as a 'distribution architecture', whereby movement is organized into sequenced flows through protocols of 'store and forward'. The multiple and sequenced sorting of people, objects and information renders the airport a highly unstable, but largely invisible, 'bioinformatic hybrid'. It is when sycnhronization fails (e.g. the baggage system breaks down) that the store and forward systems become visible, together with their architecture of anticipation. Fuller approaches this architecture from the different temporal perspectives of queuing, bringing together complex systems and the bodies of travellers in a novel perspective of contemporary airports. And while drawing on the philosophy of time, movement and becoming, she never loses sight of the sheer materiality of airports. By putting time back into the logistics of movement, Fuller is able to reveal its hidden meanings, or rather, how the architecture of store and forward contributes to the obliteration of meaning from air travel.

In the following chapter, 'Connecting the world: Analyzing global city networks through airline flows', **Ben Derudder**, **Nathalie Van Nuffel** and **Frank Witlox** examine the material spatiality of the global city network (GCN) from the perspective of airline flows. The use of airline data to illustrate and/or explain globalization is not new. There is a strong assumption that patterns of global aeromobility can help explain certain aspects of globalization. After all, if the latter is conceptualized as increasing flows (of people, goods, information, etc.) between countries, regions and cities, air travel between these is thought to provide strong evidence not only of such flows actually occurring, but also their distinctive patterns. This approach is particularly suited to approaches to globalization that focus on the GCN. In their chapter, Derudder, Van Nuffel and Witlox take one big step forward by providing a comprehensive analysis of the shortcomings of previous airline-based GCN studies. In doing so, and in previewing studies based on recently available data on airline flows, the authors provide an important analysis of the relationship between airline flows and the spatiality of GCNs. There are important methodological and theoretical issues involved in the use of airline data. Destination airports need to be distinguished from major transit ones, as the inclusion of stopovers in the data can mask the purposes of travel. And one needs to avoid the state-centrism that still besets many analyses of globalization. Moreover, new datasets allow a focus on the geography of business travel, which can be crucial for analyzing global networks. All in all, by providing a thorough conceptual and methodological critique of the use of airline data, Derudder, Van Nuffel and Witlox demonstrate the potential benefits as well as pitfalls of linking air travel to studies of globalization.

In 'Airport code/spaces', **Rob Kitchin** and **Martin Dodge** examine air travel as mediated by software and information systems. Airports, from this perspective, can be seen as 'complex assemblages' dependent on a myriad of software systems to function. By conceptualizing airport spaces as coded spaces, or code/space, that is, spaces produced by software, Kitchin and Dodge shed light upon a dimension of air travel that very few of us are familiar with and that engenders particular forms of information and surveillance. However, the authors state that, despite the pervasiveness of code/space in structuring the spatiality of airports and air travel, the work of software is always embodied, and emerges continuously through individual performances. At the same time, code/space is not a unitary space produced to a master plan. It has developed through the interlocking of different assemblages, standards and systems. All of this demonstrates that the code/spaces of air travel are relational spaces, emerging from the interplay between people and code, that is, they are performative and negotiated. Using ethnographic data, Kitchin and Dodge analyze in detail the code/spaces of check-in, security screening and immigration control, revealing the interactions between software, information and individual performance in the production of airspaces. Rather than indeterminate non-places, airports are thus seen as evolving assemblages of code/space mediated by embodied action. By focusing on the contingent and negotiated nature of the automated management of airspaces, Kitchin and Dodge make a powerful case for ethnographic studies of air travel.

Lucy Budd's 'Air craft: Producing UK airspace' provides a groundbreaking perspective on the production of airspaces. As I mentioned before, the vast majority of social research into airspaces has focused on the spaces of airports. However, contemporary aeromobilities would have not been possible at all without the production of distinctive spaces for flying, or flightpaths. Budd's chapter draws our attention to these under-researched, and mostly taken for granted, levels of airspace. Far from being mere conduits for aircraft, or abstract 'spaces of flow', airspaces are socially produced, maintained and contested. Why the sky is configured in particular ways, together with their historical development, are crucial issues for aeromobilities research. Budd takes us on a journey through the historical constitution of airspace legislation from the beginning of the twentieth century, before describing in more detail the geographies of the United Kingdom's airspace. She then looks in more detail into the ways in which these airspaces are managed, monitored and mediated by air traffic control (ATC) and piloting practices, before showing that they are also contested from the ground by anti-airport protesters. Practices of ordering, controlling, navigating and contesting the sky are all implicated in the production of the various geographies of airspace. Budd's analyses are among the first to enquire into these geographies, showing that, far from being simple neutral 'tunnels' for mobility in the sky, airspaces consist first and foremost of *social* spaces. And, as a corollary, the expansion of air travel is not solely dependent on technical and/or safety issues (e.g. how to manage airspace in order to provide the fastest and safest route from A to B). It is also contingent on political actions and decisions. The more congested the sky gets, the more contested the geographies of airspace will become.

In 'Around the world in 80 airports', **Ross Rudesch Harley** draws on Jules Verne's famous adventure in order to provide a definitive photo-essay on contemporary aeromobilities. The questions of how to describe, explain, interpret and/or represent air travel lie at the centre of the constitution of aeromobilities research as a distinctive academic field. And the almost con-sensual conclusion seems to be that traditional discursive and representational strategies of the human and social sciences will not suffice. Although this may be the case regarding mobilities research in general (and perhaps even any object of scientific investigation), aeromobilities intimate a (social) world that is fundamentally three-dimensional, and that is also essentially here *and* there at the same time. Social theory and research will definitely require the use of modes of representation that compensate for its traditional discursive shortcomings if it is to make effective inroads into the world of aeromobilities. They will include photography, moving images, graphics, sound and anima-tion. This will not only facilitate data gathering and analysis, but also inform our own representations of air travel. Harley provides us in his chapter with a powerful rendition of aeromobilities through innovative uses of photography and graphics. He invites us to imagine a journey around the world in which one never leaves the airport. Based on this very simple and intuitive principle, Harley is then able to represent the flows and materialities of airspaces with distinctive precision, representing by way of images what is, essentially, the ultimate time-space compression. With historical and theoretical textual references appended to images of both desolate and lived airspaces, Harley's photo-essay intimates a pure architecture of movement, culminating in his innovative graphical representations of air travel as a 'techno-rhizome'. Benefiting from rare access behind the airport terminal, Harley provides a unique, but at the same time comprehensive, representation of airspaces, demonstrating the centrality of visual media and art in aeromobilities research.

The third part of the book, 'The social life of air travel', develops in more explicit ways what is already intimated in the previous part: airspaces are not simply products of technical problem-solving, but are already, and after all, lived, inhabited spaces. If, on the one hand, aeromobilities research cannot proceed without profound and multifaceted analyses of the architectures of airports, the virtuality of air travel (that includes not only online booking systems, but also databases containing personal identification data) and the technology of aviation, on the other hand it must not forget that mediating those spaces are social relations of work, family life, leisure and a variety of other fields. These social relations both enable and are enabled by air travel. Whether it is the work done in airline control systems, the consumption of airport spaces, or jobs carried out while in airspaces, or even the forming of social ties across distances, the social life of air travel is a fundamental dimension of aeromobilities, even though these relations themselves cannot be understood without the mediation of the technologies and environments of air travel. Without losing sight of the complexity of aeromobilities, an analytical focus on those social relations provides invaluable insights for our emerging field. As a whole, the three chapters that compose this part of the book collectively attempt to bring out the times of airspaces, highlighting the social and human aspects of the complex systems of aeromobilities.

In 'Airborne on time', **Peter Peters** precisely veers away from studies of airports as non-places by analyzing air travel from the very opposite perspective, as workplaces. A great deal of work is necessary in order to enable reliable global air travel, and Peters, based on ethnographic research conducted at the KLM Operations Control Centre and other locations at Schiphol airport in Amsterdam (which control, among other things, the daily operations of the airline's global connections), sets out to analyze such work. Peters's focus is on how air times are produced and, as a result, on how the airline can prevent gridlocks arising in its system. Peters portrays and analyzes the work of air travel as a complex production of a heterogeneous time-space, with various systems in place for 'repairing' failures in the smooth passages of people and aeroplanes through the different sequences of air travel. Far from straightforward protocols and systems, the work of aeromobilities is also characterized by the anticipation of potential problems and disruptions. Using a series of administrative systems and technologies of representation of air travel, the various KLM departments that manage the airline's global operations create a number of practices of anticipation, evaluation and decision-making that keep those operations running as smoothly as possible. By relying on extensive ethnographic work, Peters provides a sophisticated analysis of the temporalities of situated actions and improvisations that recreate air travel's heterogeneous temporal-spatial order in real time. Using the concept of *exchange* to analyze such embodied agency, Peters demonstrates the contingent, contextualized and negotiated nature of human actions that are ultimately responsible for regular and reliable international travel.

The next chapter, 'A life in corridors: Social perspectives on aeromobilities and work in knowledge organizations', by **Claus Lassen**, contributes to bridging the gap that exists between studies of aeromobilities and those of work, family, leisure and personal relationships. Following Ogburn's classical studies, it is crucial for aeromobilities research to establish the real impact of air travel on social life, and Lassen's chapter is one of the first attempts to analyze diverse forms of social relations in the light of aeromobility. Lassen's main focus is work-related travel that takes place on a global scale and that is an increasingly dominant feature of the knowledge economy. Based on studies of two Danish knowledge organizations, Lassen conceptualizes work-related aeromobility in terms of a logic of 'corridors', a system of channels organizing contemporary social practices. The 'life in corridors' of aeromobility enables a series of social practices, from face-to-face interaction and networking over distances, to the constitution of cosmopolitan identities. Lassen identifies a number of strategies used by those performing work-related air travel, depending on whether careers or families are given prominence. Lassen's analyses are important in that they open up research into aeromobilities to individual life strategies.

In 'Getting into the flow', **Peter Adey** cuts through the surface of airport geographies in order to present them as an animate terrain constituted by practice, performance and social life. By doing that, he is able to analyze how the 'movement time-space' of airports informs identities, sensations and emotions. The flows of aeromobilities must not be seen as mere movement occurring in a containerized time-space. In order to understand the lived, sensed and experienced nature of airports in particular, Adey looks at various discourses and representations of airports, as well as ways by which people get into and out of these flows, including an innovative analysis of the quiet spaces provided inside airports for various purposes. Adey's chapter is an important contribution to debates about forms of representation of aeromobilities. He contends that issues of representation are intrinsically related to the methodology of aeromobilities research. To most observers, most of the time, airports may appear and experienced as chaotic structures, posing the question of how we can represent them and, more importantly, people's experiences of the many flows that characterize these spaces. Adey's philosophy of airport flows is grounded on extensive ethnographic research and demonstrates the difficulties of conducting research into the fluidities of aeromobility.

The final part of the book, 'Governing air travel', builds upon Kesselring's agenda for studying the politics of aeromobilities. For decades, aeromobilities have been regarded mostly as technological and economic issues. From the architecture of airports to air traffic control, technical solutions for technical problems have been the mainstay of air travel, even when safety issues are at stake. Although early aviation did flirt with both nationalism and glamorous life-styles, modern regulatory (or deregulatory) frameworks have been characterized by providing the best environment for the expansion of air travel through competition and the free market. The production, management, control and regulation of aeromobilities thus became the objects of a small number of industrialists, bureaucrats, operators and technicians. For a great deal of the history of aviation, its negative local and global impacts remained hidden. However, the last few decades have seen increasing concerns about such impacts and the development of a new politics of aeromobilities. This multifaceted politics aims at devolving decision-making about aeromobility issues from unaccountable organizations and institutions (such as central civil aviation authorities) to newly formed governance structures that are more responsive to new voices, from local groups campaigning against aircraft noise to global environmental coalitions seeking curbs to the level of carbon emissions of the aviation industry as a whole.

In 'Science, expertise, and local knowledge in airport conflicts: Towards a cosmopolitical approach', **Guillaume Faburel** and **Lisa Levy** examine a key issue for social scientists: the production of knowledge and expertise about aviation matters. Focusing on representations of impacts of airport noise on their surrounding areas, and drawing on extensive research conducted in several international airports in Europe, North America and Australia, the authors identify forms of knowledge that are local and practical, produced especially by those who live in the vicinity of airports, and that have, in the past twenty or so years, challenged the hegemonic discourses of the aviation industry. With their critiques of the limitations of a techno-scientific model for dealing with airport conflicts, territorially based stakeholders open up the field of aeromobility to local knowledge, practices and actions that inform the experiences of many of those affected negatively by air travel. As environmental issues place new stakes in the future of air travel, Faburel and Levy argue for the recognition of the local contexts and dimensions of aeromobility, while highlighting a fundamental dimension of the emerging politics of air travel.

Finally, in 'Helipads, heliports and urban air space: Governing the contested infrastructure of helicopter travel', **Saulo Cwerner** explores a less-known fact of aeromobilities. Based on research conducted in São Paulo, Brazil, a city that has experienced a massive growth in urban helicopter traffic over recent years, Cwerner examines the impact of the infrastructure of urban air travel upon the city and its residents. With particular emphasis on the system of urban heliports and rooftop helipads, which enable door-to-door aerial transportation in one of the world's largest metropolises, Cwerner analyzes the various modalities of the urban and local politics that have tried to claim urban airspace issues as part of their jurisdiction. The impact of helicopter travel upon the city is arguably far greater than the purposes and the constituency that it serves, and issues of aircraft noise and public safety are entering the public discourses of local authorities and residents' groups alike. Cwerner argues that the politics of urban aeromobilities that ensues represents an important illustration of a much more widespread politicization of airspace: a process through which airspaces become contested as the objects of democratic politics, direct action and participation.

Aeromobilities showcases a number of innovative social perspectives on aviation and air travel. It is only by bringing together geographers, sociologists, political scientists, semioticians and even artists, among other professionals in the humanities and social sciences, that aeromobilities research can develop into a major field of social enquiry. The study of the time/spaces of aeromobilities, and the ways through which they have recast social life, is a collective effort that will involve a number of theoretical perspectives and methodological endeavours. This book provides one of the first ever forums for this dialogue and exchange of ideas, and will hopefully inspire others to undertake research into a world that says as much about key issues of our time, from globalization and the environment to social networks and tourism, as any other aspect of contemporary social life.

Notes

1   Source: IATA (2007).
2   Source: Datafolha (see http://datafolha.folha.uol.com.br/po/ver_po.php?session= 480).
3   Here used as a concept – not to be confused with more specific aerial access technology of platforms and aerial lifts, which in fact do create their own particular aeromobility spaces while solving specific engineering problems.
4   Source: Department of Transport (available at www.dft.gov.uk/pgr/statistics/ datatablespublications/trsnstatsatt/publicexperiencesofandattitu1824?page=3, last accessed 2 December 2007). Figures from surveys conducted by the Civil Aviation Authority (CAA) suggest different numbers, but that is believed to derive from different sample coverage.
5   Source: Synovate (available at www.bizcommunity.com/PressOffice/PressRelease. aspx?i=478&ai=6606#contact, last accessed 2 December 2007).
6   See note 5, above.

Bibliography

Adey, P. (2004) 'Secured and sorted mobilities: examples from the airport', *Surveillance and Society*, 1: 500–19.
Airbus (2006) *Global market forecast: the future of flying 2006–2020*, Blagnac: Airbus.

ATAG (2005) *The economic & social benefits of air transport*, Geneva: Air Transport Action Group. Available at www.atag.org/files/Soceconomic-121116A.pdf (accessed 29 November 2007).

Bennett, S. (2001) *Human error – by design?* Basingstoke: Palgrave Macmillan.

—— (2006) *A sociology of commercial flight crew*, Aldershot: Ashgate.

Bilstein, R. E. (2001) *Flight in America: from the Wrights to the astronauts*, Baltimore: The John Hopkins University Press.

Boldrick, S. (2007) 'Reviewing the aerial view', *arq: Architectural Research Quarterly* 11(1): 11–15.

Boon, B. H. and Wit, R. C. N. (2005) *The contribution of aviation to the economy: assessment of arguments put forward*, Delft: CE Delft.

Bor, R. (ed.) (2003) *Passenger behaviour*, Aldershot: Ashgate.

Byman, D. L. and Waxman, M. C. (2000) 'Kosovo and the great air power debate', *International Security*, 24(4): 5–38.

CAA (2004) *CAA passenger survey report 2003*, London: Civil Aviation Authority.

COFAR (2001) *Airport city and regional embeddedness*, Paris: IAURIF.

Corn, J. J. (2002) *The winged gospel: America's romance with aviation*, Baltimore: The John Hopkins University Press.

Cornelissen, S. (2006) *The global tourism system: governance, development and lessons from South Africa*, Aldershot: Ashgate.

Cresswell, T. (2006) *On the move*, London: Routledge.

Davidson, K. (2006) 'Alternative India: transgressive spaces', in A. Jarowski and A. Pritchard (eds.) *Discourse, Communication and Tourism*, Clevedon: Channel View Publications.

Department for Transport (2003) *The future of air transport*, London: Department for Transport.

Dobson, A. (1991) *Peaceful air warfare: United States, Britain and the politics of international aviation*, Oxford: Clarendon Press.

Douglas, D. G. (1990) 'United States Women in Aviation 1940–1985', *Smithsonian Studies in Air and Space* 7, Washington: Smithsonian Institution Press.

Ebaugh, H. R. F. and Chafetz, J. S. (2000) 'Is the past prologue to the future', in H. R. F. Ebaugh and J. S. Chafetz (eds.) *Religion and the new immigrants: continuities and adaptations in immigrant congregations*, Lanham: Rowman and Littlefield.

Forsyth, P. (2006) 'Martin Kunz memorial lecture. Tourism benefits and aviation policy', *Journal of Air Transport Management*, 12(1): 3–13.

Fritzsche, P. (1992) *A nation of flyers: German aviation and popular imagination*, Cambridge: Harvard University Press.

Fuller, G. (2002) 'The arrow – directional semiotics: wayfinding in transit', *Social Semiotics* 12(3): 131–44.

—— and Harley, R. R. (2005) *Aviopolis: a book about airports*, London: Black Dog Publishing.

Gordon, A. (2004) *Naked airport: a cultural history of the world's most revolutionary structure*, New York: Metropolitan Books.

Gottdiener, M. (2001) *Life in the air: surviving the new culture of air travel*, Lanham: Rowman and Littlefield.

Griggs, S. and Howarth, D. (2006) 'Airport governance, politics and protest networks', in M. Marcussen and J. Torfing (eds.) *Democratic Network Governance Networks in Europe*, Basingstoke: Palgrave.

Grosscup, B. (2006) *Strategic terror: the politics and ethics of aerial bombardment*, London: Zed Books.

Harris, D. and Muir, H. C. (eds.) (2005) *Contemporary issues in human factors and aviation safety*, Aldershot: Ashgate.

Hochschild, A. (1983) *The managed heart*, Berkeley: The University of California Press.

Hudson, K. (1972) *Air travel: a social history*, Totowa: Rowan and Littlefield.

IATA (2007) 'IATA economic briefing: passenger and freight forecasts 2007 to 2011'. Available at www.iata.org/NR/rdonlyres/E0EEDB73-EA00-494E-9408-2B83AFF 33A7D/0/traffic_forecast_2007_2011.pdf (accessed 1 December 2007).

Kasarda, J. D. (2006) 'The rise of the aerotropolis', *The Next American City* 10.

Larsen, J., Urry, J. and Axhausen, K. (2006) *Mobilities, networks, geographies*, Aldershot: Ashgate.

Lassen, C. (2006) 'Aeromobility and work', *Environment and Planning A* 38(2): 301–12.

Law, J. (2002) *Aircraft stories: decentering the object in technosceince*, Durham and London: Duke University Press.

Lloyd, J. (2003) 'Airport technology, travel and consumption', *Space and Culture* 6(2): 93–109.

Lyon, D. (2006) 'Airport screening, surveillance and social sorting: Canadian responses to 9/11 in context', *Canadian Journal of Criminology and Criminal Justice*, 48(3): 397–411.

McPeak, M. A. and Pape, R. A. (2004) 'Hit or miss', *Foreign Affairs* 83(5): 160–3.

Mason, K. J. and Alamdari, F. (2007) 'EU network carriers, low cost carriers and consumer behaviour: a Delphi study of future trends', *Journal of Air Transport Management*, 13(5): 299–310.

Moore, M. D. (2006) 'The third wave of aeronautics: on-demand mobility'. Available at www.cafefoundation.org/v2/pdf_pav_tech/ThirdWaveTech.pdf (accessed 29 November 2007).

Ogburn, W. F., Adams, J. L. and Gilfillan, S. C. (1946) *The Social effects of aviation*, Boston: Houghton Mifflin Co.

Pape, R. A. (2004) 'The true worth of air power', *Foreign Affairs*, 83(2): 116–30.

Pascoe, D. (2001) *Airspaces*, London: Reaktion Books.

Quinn, T. (2003) *Tales from the golden age of air travel*, London: Aurum Press.

Richter, M. N. (1991) 'After forty-five years: Ogburn's predictions concerning aviation re-examined', *Technology in Society*, 13(3): 317–25.

Shaw, S. and Thomas, C. (2006) 'Social and cultural dimensions of air travel demand: hyper-mobility in the UK?', *Journal of Sustainable Tourism* 14(2): 209–15.

Sheller, M. and Urry, J. (2006) 'The new mobilities paradigm', *Environment and Planning A* 38(2): 207–26.

Upham, J., Maughan J., Raper, D. and Thomas, C. (eds.) (2003) *Towards sustainable aviation*, London: Earthscan Publications.

Urry, J. (2000) *Sociology beyond societies: mobilities for the twenty-first century*, London: Routledge.

—— (2007) *Mobilities*, Cambridge: Polity.

Whitelegg, J. (2003) *The economics of aviation: a North West England perspective*, Leyland: CPRE North West Regional Group.

—— and Cambridge, H. (2004) *Aviation and Sustainability*, Stockholm: Stockholm Environment Institute.

Wilkinson, P. and Jenkins, B. (eds.) (1999) *Aviation, terrorism and security*, London: Frank Cass Publishers.

Wohl, R. (2005) *The spectacle of flight: aviation and the western imagination 1920–1950*, New Haven: Yale University Press.

York Aviation (2004) *The social and economic impact of airports in Europe*. Airports Council International – Europe.

Part I

Aeromobilities, globalization and social theory

1 Aeromobilities and the global

John Urry

Today my favourite kind of atmosphere is the airport atmosphere . . . Airplaces and airports have my favourite kind of food service, my favourite kind of bathrooms, my favourite peppermint Life Savers, my favourite kinds of enter-tainment, my favourite loudspeaker address systems, my favourite conveyor belts, my favourite graphics and colors, the best security checks, the best views, the best perfume shops, the best employees, and the best optimism.

(Andy Warhol 1976: 145)

Airworld is a nation within a nation, with its own language, architecture, mood, and even its own currency – the token economy of airline bonus miles that I've come to value more than dollars.

(Walter Kirn 2001: 5)

Introduction

Monumental terminals of glass and steel designed by celebrity architects, gigantic planes, contested runway developments, flights massively cheaper than surface travel, new systems of 'security' – these are icons of the new global order. They are points of entry into a world of apparent hypermobility, time-space compression *and* distantiation, and the contested placing of people, cities and whole societies upon the global map. There are many ways in which flights, aeroplanes, airports and airport cities are central to an emergent global order. Without the rapid development of the complex extended systems of mass air travel, 'globalization' would be utterly different, indeed possibly it would simply never have developed in anything like the present, high-carbon form.

In this chapter I consider this possible counterfactual and examine the emergence of air travel and especially air systems. It is shown how they are central to the making of the new global dis/order. It is also shown that, not only do passengers increasingly fly across borders, forming new social time-distantiated social networks but the systems that make possible their travel also fly around and land in many towns and cities that become subject to 'airspace makeover'. Air travel, although still the practice of a very small minority of the world's population in any one year, is shown to be implicated in the global remaking of places and the contingent securing of mobile populations in a world of global riskiness.

Flying commenced at the very beginning of the twentieth century with the Wright brothers' first flight in 1903. It symbolically culminated at the end of that century with September 11 2001, when planes functioned as central 'actants' in a deadly network of spectacular globally watched destruction.[1] Air travel went from small beginnings on a sand dune in North Carolina to become the industry that stands for and represents the new global order. The history of flying is a remarkable history of various ingenious ways of transcending two-dimensional spatial constraints. It now produces an exceptional sorting and resorting of populations and places, both within countries but especially around the world. Geographical proximity in most countries no longer shapes social relationships (Castells 2001: 126), and this is in part because some people, sometimes rapidly, fly from, over and past such spatial proximities, forming new time-distantiated proximities in so doing.

In the next section I set out a brief periodization of flying and especially of airfields. Following this I examine some of the risks and systems that make possible the extraordinary achievement of 'flying' in massive machines through the air. I turn then to ten global lines of flight, before providing a brief conclusion (for further detail on much of this, see Urry (2007: chap. 7), and the various chapters in this volume).

A brief periodization

Aircraft are nothing without airspaces, and airspaces are nothing without 'the impeccable machine making use of its splendid expanses' (Pascoe 2001: 21). They are indissolubly linked, in a complex relationality with each other (see Adey (2006: 87–8)). And the history of flight has been the history of massive transformations in the 'splendid expanses' of such airspaces, of fields, runways and terminals over the twentieth century. Airspaces shift from airfield, to transport hub, and then to global hub (see Kesselring, this volume, Chapter 2).

First, early airfields in the beginning of the twentieth century were places of spectacle, record making and voyeurship. There were frequent crashes and risks for those flying and for those watching on the ground (Pascoe 2001: chap. 1; Perrow 1999: chap. 5). Initially, various inventors developed individual flying machines, machines often setting new records, akin to how cars developed as individual speed machines (Urry 2007: chap. 6). These new flying machines were astonishingly clever in leaving the airfield for short periods, so transcending the physical two-dimensional limits of land-based movement. According to Le Corbusier, aircraft were the greatest sign of progress that had so far been seen during the twentieth century, although at first the airfields did nothing to reflect this modernity (Pascoe 2001: 127). Simultaneously, flying machines transformed the nature of warfare as the new realm of airpower developed, including especially the innovation of machines that were able to bomb from the air (Pascoe 2001: 127; Kaplan 2006). But for all this there was nothing inevitable about this development of air travel. As late as the end of World War I, the Manchester Guardian authoritatively stated that aviation was a 'passing fad that would never catch on' (quoted in Thomas (2002: 3)).

However, the airfield stage had already been initiated by 1914 with the first ever commercial flight. Passengers paid $5 to fly for eight miles with the St Petersburg–Tampa Airboat Line in the US (Morrison and Winston 1995: 3), while the first international scheduled services took place in 1919 from Hounslow in west London, close to where Heathrow is now located (Pascoe 2001: 81). Early airports were oriented around the flying machines. They were mono-modal and concerned with transporting people and goods from one place to another by air. Aviation-related activities defined the meaning of these airspaces, with other activities playing a minor role. There was more or less nothing of the contemporary 'terminal' and its multiple buildings and complex intersecting activities (see Fuller (Chapter 3) and Harley (Chapter 7), both this volume).

In the second period, airports developed into transport hubs with the increasing interconnection between different modes of travel (planes, trains, metro and cars). Le Corbusier especially promoted the airport as a machine for travellers rather than as a field that is oriented to the plane (Pascoe 2001: 120–1). This quantum leap involved airspaces being turned into complex and integrated infrastructures, often with a futuristic design (Jarach 2001: 121). The airport was no longer isolated and specialised but developed into a multimodal hub so that passengers: 'are given within airport boundaries the chance to connect in a seamless way from air to ground, railway and sea ferry' (Jarach 2001: 121). However, the core business of such airspaces still involved managing the complex logistic services involved in the boarding and the deboarding of people and objects; increasingly these processes were governed by the concept of 'turnaround' time and the need for systems to minimize such time (Pascoe 2001: 125; Peters (Chapter 8), this volume). Such multimodal hubs were typically owned by public bodies, and there was a close relationship between those publicly owned airports and national carriers that were also often owned by the state. Such airports and airlines often arose out of military facilities in the era of 'organized capitalism' in which national transport interests and their intermodalities were planned and implemented by the national state (see Lash and Urry (1987), on such 'organized capitalism').

The third stage involves the further quantum leap from the traditional to the 'commercial airport', or what I would term the global hub (Jarach 2001: 123; see Kesselring (Chapter 2), this volume). Airports move away from being mainly transport hubs and become sites for mass travel. Airports are increasingly built on the edge of cities, as places or camps of banishment (Serres 1995: 19). They develop into small-scale global cities in their own right, places to meet and do business, to sustain family life and friendship, and to act as a site for liminal consumption less constrained by prescribed household income and expenditure patterns. Such airports are variably organized: through vertical public management (such as Munich), horizontal public management (such as Manchester), private–public management (such as Düsseldorf) and private corporation (such as Heathrow). But in all cases airports develop as strategically important within the global competition of places, cities and regions (see Kesselring (Chapter 2), this volume). Certain airport operators such as the Schiphol Group in the Netherlands, Fraport in Germany or BAA in the UK operate on a global scale, establishing and managing new airports and sets of airport services around the world.

Airspaces in this third period are full of commercial and tourist services for passengers, visitors and the thousands of employees. These services include bars, cafés, restaurants, and hotels, business centres, chapels and churches, shopping centres, discotheques (Munich, Frankfurt), massage centres (Changi), conference centres (Munich), art galleries (Schiphol), gyms (Los Angeles) and casinos (Schiphol). Also, there are many instant offices and airport hotels, allowing travellers to arrive, stay over, do their 'business' face to face and depart, especially as many airports are now organized on the hub-and-spoke model (Doyle and Nathan 2001). Airspaces are thus places of 'meetingness' that transforms them into strategic moments in constructing a global order. They develop into 'destinations' in their own right, ambivalent places, of multiple forms of transport, commerce, entertainment, experience, meetings and events. Such airspaces provide multiple sites for developing and sustaining what Knorr Cetina describes as 'global microstructures' (2005).

Risks and systems

Central to all air travel is risk. Large technical systems such as airports and the global aviation industry are sites of riskiness, and this has been so since early planes ventured into the sky a century ago. Flight is risky for those flying, for those organizing and managing those flights, and for those on the ground as viewers or innocent bystanders. The riskiness of air travel is of course the stuff of novels and movies, such as Arthur Hailey's *Airport* or Michael Crichton's *Airframe*, which have explored many of the possible risks in prescient detail. A wide array of software-based 'expert systems' have been developed to deal with this riskiness and contingency of air travel. These have remarkably transformed the hazards and physicality of taking off, cruising and landing. Various non-human actants and, especially, much computer code are combined with rule-following humans to enable this wonder of relatively safe mass air travel. Air traffic control systems effect high levels of safe take-offs and landings, while the Boeing 777 contains some seventy-nine different computer systems requiring 4 million lines of code (Dodge and Kitchin 2004: 201; Kitchin and Dodge (Chapter 5), this volume). Indeed if, as Thrift and French argue, software conditions existence in the city, this is so in a deeper and more extensive form within airspaces (2002; Adey and Bevan 2006). There is no airspace and there are no smoothly flying citizens without vast amounts of computer code. There has been the development of pervasive, consistent and routinized 'code/space' producing the 'real virtuality of air travel' (Dodge and Kitchin 2004). In these various code/spaces, the code is so significant that without code the space fails. There is no alternative to the code, even though it typically remains in the background and unnoticed until, of course, the systems crash (Adey 2006: 80). The airline passenger ticket is the material embodiment of such code/space, on which are printed several data codes that describe what the passenger is doing but also simulate and predict other actions each passenger may undertake (Kitchin and Dodge (Chapter 5), this volume).

The extremely complex management of multiple movements at airports involves various expert systems. These computer software systems, which have developed in a piecemeal way, are intended to orchestrate take-offs, landings, ticketing and reservations, baggage handling, schedules, cleaning, weather forecasting, in-flight catering, security, multiple employment patterns, baggage X-ray, waste management, environmental impact and so on (there are twelve such systems at Schiphol (Peters 2006: 115)).

Such systems have built into them risk assessments of events such as delayed flights, sick crew, damaged planes, adverse weather, computer crashes, terrorist bombings and so on. The central resources involved in managing airflights are time, money and capacity. Peters shows how continuous modifications and adjustments are made between these different resources in order to keep the system moving and especially to get the planes 'airborne on time' (Peters 2006: 122–4). Indeed, in the third period air travel has developed into a globally competitive industry (or set of industries) since the previously organized capitalism with protected national flag carriers has mostly dissolved. Through so-called 'open skies' policies, there is enhanced global competition, there is 'disorganized capitalism' (Lash and Urry 1987, 1994). This engenders a massive push to minimize the turnaround time of planes and of their crews. Many interrelated events must be synchronized so as to minimize the periods in which planes and the crew rest unproductively on the ground. Significant changes in this have been especially initiated in the last decade or two by the budget airlines of North America, Western Europe and now India and China. These airlines have increased profit margins per passenger by reducing turnaround times, developed exclusive internet bookings, deployed demand-responsive pricing, used cheaper airports, simplified check-in procedures and reduced labour costs often through non-union workplaces. These budget airlines mostly fly point-to-point, and this has also led to some reduction in the number of systems that have to be synchronized (such as planes needing to wait for connecting flights).

In general, this tight coupling of such interactively complex systems makes airlines and airports especially vulnerable to small disruptions that produce cascading and positive feedback effects when things go slightly wrong. Complex machine systems create big accidents when systems malfunction, unlike the small accidents when a walker falls, a horse dies or a coach over-turns. The powerful system can crash when one small part of it malfunctions.

Perrow specifically dissects various aircraft crashes resulting from small events, often through 'management' error, as well as many involving positive feedback loops, where 'humans' or 'systems' respond either to the wrong aircraft or to its insertion into the wrong sequence or to a false interpretation of information upon a display and so on (1999: 141, 160). Over time there have been large reductions in air travel fatalities, partly because of technological improvements and partly by introducing forms of redundancy into the systems (four engines rather than two). However, although such developments have improved air safety technically, the historic reduction in the number of fatalities and injuries has slowed down because, according to Perrow, various commercial and military demands have substantially increased (1999). Airline operators have sought greater speed, reduced 'manning' levels, higher altitude flying, reduced fuel use, greater traffic density, reduced separation between planes, and more operations to take place in all weathers. These heightened requirements have engendered a tighter coupling of the systems and an enhanced workload on the crew and on air traffic control at very specific moments, moments that push the system to its limits (Perrow 1999: 128–31, 146).

Global lines of flight

Regular and relatively risk-free air travel is centrally implicated in producing global ordering, which should be viewed as enacted as process and as multiple performances, more as effects and less as 'cause' (Urry 2003). This global space constitutes its own domains through multiple processes, including those technologies and systems that afford relatively risk-free, long-distance air travel. Air flights are central to performing the global order. Such an umbilical cord between air travel and the global can be seen first in the sheer scale of the air travel industry, international travel flows and the monumental scale of airports, all of which are major components of the emergent global economy. As noted above, travel and tourism together constitute the largest industry in the world, worth $6.5 trillion and directly and indirectly accounting for 8.7 per cent of world employment and 10.3 per cent of world GDP (World Travel and Tourism Council 2006). The 1.5 billion air journeys that take place each year occur within increasingly massive and iconic airports. Thus Beijing's Terminal-3, designed by celebrity-architect Norman Foster, possesses a: 'soaring aerodynamic roof [that] will reflect the poetry of flight. Passengers will enjoy a fully glazed single, lofty space, day lit through roof lights and bathed in colour changing from red to yellow as you progress through it'.[2] The completion of the terminal will make Beijing Capital International airport the largest in the world, outdoing both the current largest international airports, Chek Lap Kok in Hong Kong and Heathrow in the UK, although the building there of Terminal 5 may alter that again in this incessant global competition for airspaces, cities and societies (see Edwards (1998) on airport design; Harley (Chapter 7) and Kesselring (Chapter 2), both this volume).

Second, air travel presupposes the notion of a global or Universal Time. Such a time synchronizes the actions of all organizations and people involved in air travel around the world. It is thus an 'industry' presupposing a global ordering and employing a notion of time that synchronizes air flights, a notion of time involving the heterogeneous ordering of aircraft, passengers, crew, baggage, fuel, freight and catering, which have to be assembled so that planes can be 'airborne on time' (Peters 2006: chap. 5; Peters (Chapter 8), this volume). What is central to all these systems is the notion of time and especially of Universal Time that synchronizes the actions of people and countless organizations around the world (Peters 2006). Dispersed and heterogeneous flows of different categories of peoples and objects are all synchronized, and this synchronization is based upon a universal measure of time. This is necessary so that each plane in each airport in each air sector can get 'airborne on time' (or more or less on time: Peters 2006). Synchronization has to be global and based on a system of time reckoning that is also global. Such synchronization is also effected by the more or less universal interconnectedness of the multiple computerized booking systems and by the general use of English as the language of global airspaces (Peters 2006: 105).

Third, air travel is a quintessential mode of dwelling within the contemporary globalizing world, a world of arrivals and departures, lounges, duty free, English signs, pre-packaged food, frequent flyer programmes and what Andy Warhol terms 'the airport atmosphere' (1976: 145). Many iconic signs and practices associated with the global order derive from international air travel and a widespread familiarity that is especially spread through film and TV (see Iyer (2000), on the 'global soul' and Makimoto and Manners (1997), on the 'nomadic urge'). J. G. Ballard describes the symbolic aspect of the global display seen in airport concourses, which: 'are the ramblas and agoras of the future city, time-free zones where all the clocks of the world are displayed, an atlas of arrivals and destinations forever updating itself, where briefly we become true world citizens' (cited in Pascoe (2001: 34)). Moreover, airspaces teach people through contemporary 'morality plays' the appropriate categories by which to navigate the conflicts and dilemmas of the contemporary world. These categories include business-class male, terrorist, Third-Worlder, suspect Arab, Westerner, budget traveller, female service worker, illegal migrant and so on. Aaltola thus maintains that the: 'airport provides a particularly well-suited place in which to learn the hierarchical world-order imagination . . . Placed in the airport, a person recognizes the types and remembers their own respective position among them', positioned through the distinctions of class, gender, ethnicity, age and so on (2005: 275). We can note the well-trained eye of the 'airport gaze' in which since 2001 it is legitimate to be ever vigilant, since air travellers differ greatly in their perceived threats to the new world order, and increased securitization is deemed essential for dwelling in this risky global era.

Fourth, air travel transmits people into global networks through a 'space of transition' (Gottdiener 2001: 10–11). Air travel is the key 'space of flows' that moves people around the world, especially connecting together hub airports located in major 'global' cities (Castells 1996; Aaltola 2005: 267; Derudder, Van Nuffel and Witlox (Chapter 4), this volume). This system of airports is key to the constitution of global networks, permitting travel so that people encounter other people and places from around the world 'face to face'. Air flights are centrally significant microstructures within the performances involved in the global order. This *system* links together places, forming networks and bringing those connected places closer together. Two hub airports are 'near' in the network of air travel, even if they are thousands of miles apart. The 2003 Global Airport Monitor identified fifty-one hubs, twenty-five European, fourteen American, nine Asian-Pacific and three African (Aaltola 2005: 267). Simultaneously, these networks distance those other places, the spokes that are not so well connected, not part of 'hub civilizations' as Huntington puts it (1993). Thus the development of complex networks of air travel (with at least 300 airlines worldwide who are members of IATA) produces areas of very dense air traffic (between the hubs) and other areas (the spokes) characterized by sparse networks that have the effect of peripheralizing people and places away from the hubs of the global order (Graham and Marvin 2001; Derudder, Van Nuffel and Witlox (Chapter 4), this volume).

Fifth, air travel and its visible inequalities are a synecdoche of the increasingly global pattern of social inequalities deriving from huge variations in what I conceptualize as 'network capital' (see Urry 2007: chap. 9). Globalizing systems are highly differentiated by the kinds of traveller moving through these semi-public airspaces. In particular, the global or kinetic elite experiences: 'the construction of a (relatively) secluded space across the world along the connecting lines of the space of flows' (Castells 1996: 417). For first-class passengers air travel is integrally interconnected with limousines, taxis, air-conditioned offices, fast check-in and fast routing through immigration, business class hotels and restaurants, forming a seamless scape along which nomadic executives *making* the global order can with less effort travel (often needing little or no money since much food, drink, travel are free; there are 'free lunches'). For countless others, their journeys are longer, more uncertain, more risky and indicative of their global inferiority in a world where access to network capital is of major significance within the emerging global stratification system. Queueing is a key demarcator of social hierarchy (Fuller (Chapter 3), this volume). Moreover, automated software for sorting travellers as they pass through automatic surveillance systems, such as iris-recognition for Privium passengers at Schiphol, reinforces the 'kinetic elite', whose ease of mobility differentiates them from a low-speed, low-mobility, queueing mass (Wood and Graham 2006). Privium is described as 'a select way to travel', 'an exclusive membership for frequent travellers who appreciate priority, speed and comfort and who like to start their journey in style'.[3] And the enhanced mobility of this elite is at the expense of the slowing down and greater time that is thus made possible so as to interrogate those who are not part of the Privium club (Adey 2006: 89).

Sixth, airports are culturally complex places, since peoples and cultures from around the world overlap within them through the intersection of enormously elaborate relays. These relays of peoples, and to some extent objects, especially come together within departure lounges. These are paradoxical sites, places of intense sameness ('no-places') produced by the systems of the aviation industry *and* of intense hybridity as mobile peoples and cultures unpredictably intersect through various modes of 'dwelling-in-transit'. As Serres writes:

> On the departure board, the list of destinations reads like a gazetteer of the world ... Via the operations of this particular message-bearing system, men and women part company and come together, re-arrange themselves and create new human mixes. Here we see them at rest; in a short while people who are now standing next to each other will be a thousand miles apart, and strangers will converge into neighbourliness.
>
> (1995: 258)

And daily flows through airports contribute to the production of contemporary urbanism, including diasporic cultural communities, 'ethnic' restaurants and neighbourhoods, distant families and cosmopolitan identities, and exclusive zones and corridors of connectivity for the fast-tracked kinetic elite (see Lassen (Chapter 9), this volume on those 'corridors').

Seventh, international air traffic makes possible an extraordinary diversity of mobilities, of holiday-making, money laundering, business travel, drug trade, infections, international crime, food transport, asylum seeking, leisure travel, arms trading, people smuggling and slave trading (Hannam *et al.* 2006: 5–9). These intersecting mobilities produce the chaotic juxtapositions of different spaces and networks. For example, global diseases rapidly move. Thus the:

> world has rapidly become much more vulnerable to the eruption and, more critically, to the widespread and even global spread of both new and old infectious diseases ... The jet plane itself, and its cargo, can carry insects and infectious agents into new ecologic settings.
>
> (Mann, cited in Buchanan (2002: 172))

Only a few, long-range random transport connections are necessary to generate pandemics, such as occurred among those threatened by SARs. During 2003 it spread across the very mobile Chinese diaspora between south China, Hong Kong and Toronto (Sum 2004; Hannam *et al.* 2006: 7). While foot and mouth first appeared in central India in 1990, by 1995 it had spread across much of India, and by 1998 it had inserted itself into the international trade in animal products and was moving more rapidly. It appeared in Malaysia, in various impoverished countries of East Africa, and in Iran, Iraq and Turkey. By 2001 it had appeared in countries that had been free of foot and mouth, including South Korea, Japan and the UK (see Law 2006).

Eighth, air flight affords a god's eye view, a view of the earth from above, with places, towns and cities laid out as though they are a form of nature waiting for the 'possessive gaze'. Air travel generates 'map-readers' rather than 'wayfinders' (Ingold 2000). While way-finders move around *within* a world, map-readers move across a surface as imagined from above. And air travel colludes in producing and reinforcing the language of space as imagined from above, constituted of abstract mobilities and comparison, the expression of a mobile, abstracted being-in-the-world (see Budd (Chapter 6), this volume, on air space). And through this mode places become transformed into collections of abstract characteristics in a mobile world, ever easier to be visited, appreciated and compared even from above, but not really known from within (see Szerszynski and Urry (2006) on how a vision from above is engineered through the iconic pictures of the earth taken from 'space').

Ninth, air travel is indissolubly bound up with the new relations of empire that, according to various authors, increasingly replace nation-state sovereignty or 'society' (Hardt and Negri 2000; on empire and air travel, see Aaltola (2005)). By 'empire' is meant the emergence of a dynamic and flexible systemic structure articulated horizontally across the globe, a 'governance without government' that sweeps together all actors within the order as a whole (Hardt and Negri 2000: 13–4). Empire involves a system of nodes and connecting lines that is replacing the world atlas. Societies are drawn over time into the 'basin' of empire. Empire is to be understood as a 'network-based imperial hierarchy' without necessarily strong co-present territorialities (Aaltola 2005: 268). Contemporary societies increasingly possess the characteristics of empire. Societies are drawn into the attractor of empire upon the world-as-stage, competing for the best buildings, world heritage landscapes, celebrity-designed airports, global brands, skyline, palaces, galleries, stadia, infrastructures, games, sports heroes, skilled workforce, universities and security, while beyond the imperial centre effects spread across nominally distinct national borders as imperial representatives fly to the empire's spokes from the imperial hub. Thus the 'network-based imperial hierarchy is knit together by the air travel system . . . [it] manages to create an economical, vibrant and political power hub geography . . . [that] signifies a healthy, stable and predictable world order' (Aaltola 2005: 268). The US is the most powerful empire currently upon the current world-as-stage, with various exceptional centres (NY, LA, Washington), many icons of power (Pentagon, Wall Street, Hollywood, Ivy League Universities, Texan oil wells, Silicon Valley, MOMA), a dense transportation infrastructure, a porosity of certain borders, huge 'imperial' economic and social inequalities, and networks linking it through travel and communications with almost every other society. Yet each society as empire produces its opposite, a co-evolving other, its rebellious multitude with the US empire generating a powerful multitudinous 'other'. Huge transformations are taking place in the production of 'empire-and-multitude' through 'global fluids' of money laundering, drug trade, urban crime, asylum seeking, people smuggling, slave trading and urban terrorism. These and many other fluids all depend upon the passages of passengers passing through airports, and as a result the spaces of multitude and empire are contingently juxtaposed as they dramatically were on September 11 2001 (Urry 2003; 2007).

Finally, air spaces are neither non-places nor places of what is understood as conventional dwellingness. Cities are becoming more like airports, less places of specific dwellingness and more organized in and through diverse mobilities and the regulation of those multiple mobilities. Indeed, the more apparently 'cosmopolitan' the place, the more that place is produced and consumed through multiple mobilities, very much akin to the multiple ways that airports function and are organized (Sheller and Urry 2004). Also airports are themselves increasingly vast cities that may well harbinger a particular conception of future urban form (Fuller and Harley 2005). And all cities are increasingly like airports in that forms of surveillance, monitoring and regulation are being implemented as part of the global 'war on terror'. In what has been called the 'frisk society', detention centres, CCTV, GPS systems, iris-recognition security and intermodal traffic interchanges, once trialled within airports, move out to become mundane characteristics of towns and cities, places of fear and contingent ordering within the new world order. Hence, Martinotti writes that airports and the like 'are the places of the city we live in today. Non-places are nothing less that the typical places of the city of our times' (1999: 170). Thus airspaces are the future not just for those who are literally *Up in the Air* (Kirn 2001). Diken and Laustsen describe the nature of current societies: 'in which exception is the rule, a society in which the logic of the camp is generalized' (2005: 147); and my argument here is that it is global air travel and its specific 'mobilities and materialities' that are turning spaces of exception (airspaces) into the generalized rule for urban design (Sheller and Urry 2006a). Thus airspaces are typical of those 'places' that the global order is ushering in, showing many overlaps and similarities with towns and cities from around the world. It is increasingly difficult to distinguish between airspaces and other places in a global order; there is de-differentiation as systems of air travel move out and increasingly populate many kinds of place. According to Diken and Laustsen, the camp of the airspace has become the rule (2005: 147). So not only do passengers increasingly fly around the world, but the systems of both movement and securitization that make possible such travel also fly around, landing in many towns and cities. As Fuller and Harley state: 'the airport is the city of the future' (2005: 48).

Conclusion

Air flights, air systems and air-mindedness are thus central to the emergent global order (see Adey (Chapter 10), this volume). They generate mass movement, new iconic architectural forms, new forms of dwelling, interconnectedness, new inequalities, novel global meeting places, distinct ambivalent juxtapositions, new modes of vision and enhanced relations of 'empire' as attractors, new securities and systems of surveillance and new forms of protest.

And these systems do not just stay in the airport but move out and increasingly transform many other towns and cities, which become similarly remade as securitized airport-type spaces. And it may be that we have not seen anything yet if Virgin has its way, and space travel is the next frontier for global aeromobility.[4] This could well engender a fourth stage in airspaces, perhaps also with many personalized 'flying cars' (see Cwerner (Chapter 12), this volume).

However, it may also be that global heating, as Lovelock characterizes contemporary climate change, so kicks in within two or three decades that many of the world's current air systems come to be mothballed and left to rust away in the relentless heat, helped probably by global protest movements (Lovelock 2006). And if they are mothballed, much of what the late twentieth century saw as the global borderless future may also come to be mothballed. The global order may come thus to be seen as part of the long-term, unsustainable, energy-consuming hubris of the late twentieth century, and it and its extraordinary flying machines may begin to retreat really quite soon.

Two visions then for the future: Virgin Galactica and flying cars for all, or rusting planes and derelict runways as droughts and floods ravish the land. What happens to air travel will index what happens to the high-carbon societies, of which flying machines for the masses were perhaps one hubris too far.

Notes

1   See en.wikipedia.org/wiki/Wright_brothers, accessed 18 March 06, on the many controversies as to who actually 'flew' first.
2   Cited in *The Hindu*: www.hindu.com/2006/03/03/stories/2006030301412000.htm, accessed 18 March 06. China has recently announced that it plans to double air traffic, with 100 new planes a year; see www.guardian.co.uk/china/story/0,,170 9834,00.html (accessed 7 December 07).
3   See www.schiphol.nl/schiphol/privium/privium_home.jsp#anchor3, accessed 20 March 06.
4   See www.virgingalactic.com/htmlsite/book.htm, for the booking form for Virgin Galactic flights that are due to start in 2009.

Bibliography

Aaltola, M. (2005) 'The international airport: the hub-and-spoke pedagogy of the American Empire', *Global Networks*, 5: 261–78.
Adey, P. (2006) 'Airports and airmindedness: spacing, timing and using Liverpool Airport, 1929–1939', *Social and Cultural Geography*, 7: 343–63.
—— and Bevan, P. (2006) 'Between the physical and the virtual: Connected mobilities', in M. Sheller, M. and J. Urry (eds.) *Mobile Technologies of the City*, London: Routledge.
Buchanan, M. (2002) *Small world: uncovering nature's hidden networks*, London: Weidenfeld and Nicolson.
Castells, M. (1996) *The rise of the network society*, Oxford: Blackwell.

—— (2001) *The Internet Galaxy*. Oxford University Press.

Diken, B. and Laustsen, C. (2005) *The Culture of Exception. Sociology Facing the Camp*, London: Routledge.

Dodge, M. and Kitchin, R. (2004) 'Flying through code/space: the real virtuality of air travel', *Environment and Planning A,* 36: 195–211.

Doyle, J. and Nathan, M. (2001) *Wherever next: work in a mobile world*, London: The Industrial Society.

Edwards, B. (1998) *Modern Terminal: New Approaches to Airport Architecture*, New York: E&FN Spon.

Fuller, G. and Harley, R. (2005) *Aviopolis. A Book about Airports*, London: Black Dog Publishing.

Gottdiener, M. (2001) *Life in the air: surviving the new culture of air travel*, Lanham: Rowman and Littlefield.

Graham, S. and Marvin, S. (2001) *Splintering urbanism: networked infrastructures, technological mobilities and the urban condition*, London: Routledge.

Hajer, M. (1999) 'Zero-Friction Society', *Urban Design Quarterly*, 71: 29–34.

Hannam, K., Sheller, M. and Urry, J. (2006) 'Editorial: Mobilities, Immobilities and Moorings', *Mobilities*, 1: 1–22.

Hardt, M. and Negri, A. (2000) *Empire*, Cambridge, Massachusetts: Harvard University Press.

Huntington, S. P. (1993) 'The clash of civilizations?', *Foreign Affairs*, 72(3): 22–49.

Ingold, T. (2000) *The Perception of the environment: essays in livelihood, dwelling and skill*, London: Routledge.

Iyer, P. (2000) *The Global Soul*, London: Bloomsbury.

Jarach, D. (2001) 'The evolution of airport management practices: towards a multi-point, multi-service, marketing driven firm', *Journal of Air Transport Management*, 7: 119–25.

Kaplan, C. (2006) 'Mobility and war: the cosmic view of US "air power"', *Environmental and Planning A*, 38(2): 395–407.

Kirn, W. (2001) *Up in the Air*, New York: Doubleday.

Knorr Cetina, K. (2005) 'Complex global microstructures: the new terrorist societies', *Theory, Culture & Society*, 22(5): 213–34.

Lash, S. and Urry, J. (1987) *The End of Organized Capitalism*, Cambridge: Polity.

—— and —— (1994) *Economies of Signs and Space*, London: Sage.

Law, J. (2006) 'Disaster in agriculture: or foot and mouth mobilities', *Environment and Planning A*, 38: 227–39.

Lovelock, J. (2006) *The revenge of Gaia. Why is the Earth fighting back – and how we can still save humanity*, Santa Barbara, California: Allen Lane.

Makimoto, T. and Manners, D. (1997) *Digital nomads*, London: John Wiley and Sons.

Martinotti, G. (1999) 'A city for whom? Transients and public life in the second-generation metropolis', in R. Beauregard and S. Body-Gendrot (eds.) *The urban moment: cosmopolitan essays on the late-20th-century city*, London: Sage.

Morrison, S. A. and Winston, C. (1995) *The evolution of the airline industry*, Washington, DC: Brookings Institution Press.

Pascoe, D. (2001) *Airspaces*, London: Reaktion.

Perrow, C. (1999) *Normal Accidents*, Princeton: Princeton University Press.

Peters, P. (2006) *Time, Innovation and Mobilities*, London: Routledge.

Serres, M. (1995) *Angels: a modern myth*, Paris: Flammarion.

Sheller, M. and Urry, J. (eds.) (2004) *Tourism Mobilities: Places to Play, Places in Play*, London: Routledge.

—— and —— (eds) (2006a) *Mobile Technologies of the City*, London: Routledge.

—— and —— (2006b) 'The new mobilities paradigm', *Environment and Planning A*, 38: 207–26.

Sum, N.-L. (2004) 'The paradox of a tourist centre: Hong Kong as a site of play and a place of fear', in M. Sheller and J. Urry (eds.) *Tourism Mobilities: Places to Play, Places in Play*. London: Routledge.

Szerszynski, B. and Urry, J. (2006) 'Visuality, mobility and the cosmopolitan: inhabiting the world from afar', *British Journal of Sociology*, 57: 113–32.

Thomas, C. (2002) *Academic Study into the Social Effects of UK Air Travel*, London: Freedom-to-Fly.

Thrift, N. and French, S. (2002) 'The automatic production of space', *Transactions of the Institute of British Geographers New Series*, 27: 309–35.

Urry, J. (2003) *Global Complexity*, Cambridge: Polity.

—— (2007) *Mobilities*, Cambridge: Polity.

Warhol, A. (1976) *The Philosophy of Andy Warhol (From A to B and back again)*, London: Picador.

Wood, D. and Graham, S. (2006) 'Permeable boundaries in the software-sorted society: Surveillance and the differentiation of mobility', in M. Sheller and J. Urry (eds.) *Mobile Technologies of the City*, London: Routledge.

World Travel and Tourism Council (2006) *Progress and priorities 2006/7*, London: World Travel and Tourism Council.

2   Global transfer points

The making of airports in the mobile risk society

Sven Kesselring

> Understanding power in the global age needs a mobility-related research that focuses on places of flows and the power techniques and the strategies of boundary management that define and construct places and scapes where cosmopolitization is possible.
>
> (Beck 2008: 34)

Airports are fascinating places, but as objects of social-scientific mobility research they are almost entirely uncharted territory. This strikes us as all the more surprising as international airports, the transfer points of international air travel, play a fundamental role in the globalization of society and the economy (Urry 2007: 154 ff.). The French spatial theorist Henri Lefebvre saw this as early as the 1970s. He speaks of the 'geopolitics of air travel' (Lefebvre 2000: 365) and points up the structuralizing influence of global spatial mobility. The idea that transcontinental air travel networks and airport transfer points reflected a new global spatial matrix occurred to more prescient observers in the aircraft and airline industries early on:

> The marketing experts at Lockheed may have been thinking [when they christened their new airliner Super Constellation] . . . of a network of air travel routes that, like a heavenly constellation, spanned the earth and thus defined a unique space of their own. Thus it is a product name that also denotes how air travel functions to create new constellations.
>
> (Asendorf 1997: v)

Today, the network of global airline connections is emblematic of the cosmopolitanism of the modern world not only conceptually but in fact (Keeling 1995; Smith and Timberlake 1995; Taylor 2004; Derudder and Witlox 2005; Hannam *et al.* 2006; Urry 2007), and airports symbolize and embody the global paradigm of the 'mobile risk society' (Kesselring 2008; Beck 1992). Airports are essential elements of the mobility potential of that society.

Today, the world's political, economic and cultural organizations do not limit their activities to nationally defined spaces. Prime movers use the whole world as a stage for their projects and plans. This is not only to be understood in the sense of military conquest. Decision-making and follow-up action take place on a global scale, whether in the form of concerted political action or as economic decisions aimed at the world market, but even individual lifestyles presuppose decentralized and networked mobility management (see Kesselring 2006a). The precondition for this is a highly developed social mobility potential, which enables individuals, corporate entities, ideas and goods to move and interconnect globally. Airports are the interfaces between the territorial and the global spaces in which this movement takes place.

Airports point up another side of globalization too, perhaps the dominant one. As the ground units of an aerial mode of mobility, as mobility machines in the sense that modern habitation units are machines for living, airports generate new social inequities by creating a stationary auxiliary personnel for the mobility of the mobile (Adey 2006). They epitomize the immense costs of globalization in their consumption of space and in the ecological and social side effects (see Faburel 2003). Their existence goes hand in hand with the progressive modernization of transport systems connecting cities, regions and nations as well as continents (Zorn 1977).

The expansion of an airport in (sub)urban space is fraught with conflict. The controversy over the construction of a third runway at Logan International Airport in Boston has been raging for thirty years (Faburel 2003: 7). Resistance to Runway 18 West at Frankfurt International Airport persisted for over twenty years and continues to excite political passions to this day (Troost 2003; Geis 2005). The establishment of a completely new transfer point within existing networks and territorial spatial and social settings leads to even greater turbulence and social upheaval, sometimes politicizing whole regions. The paradigmatic example for the political and social role that airports and plans for airport construction can play as 'mobilization space(s)' (Apter and Sawa 1984: 7) is the long-term conflict around Narita International Airport near Tokyo. It started as a low-intensity conflict in the early 1960s, when the Japanese government declared the airport 'a symbol of the new role Japan would play in the world' (Apter and Sawa 1984: 5). In 1965 the conflict turned violent. And when the airport was opened in 1978 and got its second runway in 2002, Narita Airport had become a national symbol for the most violent civil conflict Japan had experienced since the Second World War. Business and political leaders usually equate new airport projects with positive economic development, as the case of Narita Airport shows, while the critical citizenry remains skeptical of unchecked globalization-driven airport construction. Even the seemingly straightforward practical points in airport design raise questions as to society's real need for mobility. The demand for global air mobility potentials cuts to the core of the social, geographical and cultural structures of urban and suburban spaces (Hartwig 2000; Brueckner 2003). To date, the ecological and social consequences of the air travel system have been disastrous (see www.pa.op.dlr.de/aac/; www.germanwatch.org; Lassen 2006). The concept of 'sustainable aviation' (Thomas *et al.* 2003) may thus seem oxymoronic, but it points to the political challenge arising from air travel's status as a mobility commonplace.

Social-scientific mobility research is only just beginning to enquire into the social side effects of increased air travel (see Urry 2007). It is clear that such knowledge is highly relevant to informed policy-making when decisions must be taken as to airport design and location. Airport policy is at the heart of the mobile risk society. Global issues, the parameters for the globalization of society and the economy, must be worked out and digested locally (Brenner 2004). In this sense airports are paradigmatic localities, not merely 'flow-throughs' where travelers and goods from all over the world arrive and depart, but rather the place where global and local influentials and the interests they represent interact and seek solutions. 'Expanding airports are manifestations of "glocalization". . . . They are settings for the global expansion and intensification of mobility on the one hand and for local infrastructure transformation on the other' (Geis 2005: 130).

Cities and regions are directly affected, with globalization bringing about changes in their socio-material morphology (Graham and Marvin 2001; Oswalt 2004). Globalization lands on the runways of the international airport hubs – but it also takes off from them. Air travel largely defines the transnational time and mobility regimes of world society. Thus the question of the 'glocal' character of politics (Robertson 1992; Swyngedouw 1997; Beck 2000) can be exemplified and empirically studied in airport policy-making.

Four hypotheses form the basis of the following study: First, I proceed from the assumption of a fundamental 'global shift' (Dicken 2003), a structural shift in the centers of economic and political decision-making power that has redrawn the map of the world to create dynamic 'world city networks' (Knox and Taylor 1995; Taylor 2004; Smith and Timberlake 1995; Castells 1996). Second, air travel connections between these central locations can serve as reliable indicators for global transformations (Keeling 1995; Derudder and Witlox 2005; Derudder *et al.* 2005). The structural change in mobility towards a greater social significance of 'aeromobility' points to the process of global restructuring in which most societies find themselves today (Urry 2007; Hannam *et al.* 2006). Third, airports assume the role of stabilizing units, spatial fixities, in this restructuring process. The global social, political and economic space of world society stretches out from airports, and airports sustain the uninterrupted connectivity between regions and cities across the globe. Airports connect 'spaces of globalization' with 'spaces of territoriality' (Brenner 2004: 55); thus they are 'global transfer points'. And, fourth, airports are the intersection of all regulative levels of global society. The function of airports can be described as a kind of transmission for shifting from global to local and vice versa (see Kesselring 2006a). They are thus characterized by the 'politics of scale' (Swyngedouw 1997; Brenner 1997, 2004), which simultaneously transforms them into ambivalent political settings with ambiguous logics and functional principles that have been insufficiently investigated to date (see Beck *et al.* 1999). Airport politics thus represent a more general transformation in politics in mobile risk society, as societies do not know how to manage and enclose the conflictual dynamics that go along with the worldwide deployment of global airborne mobility potentials.

Structural changes in mobility, or: stories air travel tells of the mobile risk society

In the mid 1970s, the old Tokyo Haneda Airport was of regional importance at best, with direct flights only to other Japanese cities and to the nearby Asian mainland (Keeling 1995: 119). Needless to say, this has changed. In 2006 Narita International Airport handled 31.8 million passengers.[1] This alone shows that the Japanese capital is one of the world's 'global cities' (Sassen 1991), its airport a global transfer point for passengers and goods from all over the world. In the geopolitics of air travel, Tokyo rates as an 'alpha world city', along with Frankfurt, Hong Kong, London, Milan, New York, Paris and Singapore (O'Connor 2003: 91). Tokyo illustrates the profound structural changes in spatial mobility in the global era. The configuration of the world-wide network of air travel routes and airports shows the rise and fall of urban centers in the hierarchy of global society.

The close connection between the route map importance and the geopolitical importance of airports and cities amounts to a political geography of globalization. In 2000, the airports in four city-regions – London, New York, Chicago and Tokyo – accounted for 23 per cent of total passenger movement through the world's 100 busiest airports (see O'Connor 2003: 90). The route map centrality of an airport correlates with the economic and political potency of cities and regions. As Keeling (1995: 119) noted in the mid 1990s,

> for cities and regions a non-stop flight to London is a direct pipeline into the world economy. . . . A map of international air connections clearly illustrates the major global linkages between New York, London, and Tokyo, and the role the cities play as dominant global hubs.

The importance of air transport should not be measured in terms of sheer tonnage; more than 90 per cent of the cross-border trade in goods is still handled by ship (Gerstenberger and Welke 2002; Rodrigue *et al.* 2005). But the economy and important segments of society increasingly depend on air travel and transport. More and more high-value-added goods and foodstuffs are being transported by air. This gives rise to economic structures and inter-dependencies that mesh with the schedules of airline networks. In particular, the business elite travels by air. Key managers and key employees such as facility start-up engineers are in the air up to 150 days a year and are sometimes responsible for territories covering several continents: Europe, Mideast and Africa (EMEA) or the US, Latin America and the Caribbean. Such geographical reach is based on the clear assumption that airports are the open gateways to the world. Key people are not bound to nations and continents, and the sum of flight connections that can be made at any one locality defines the reach of that locality's economic, political and cultural influentials.

The consequence of such mobility for the traveler is a kind of permanent emotional and psychological transition state:

> Brussels on Monday evening directly to the hotel, a meeting with the European executives, dinner with them, a beer in the bar, and then to bed. The next day in Belgium at a strategic meeting that lasted all day till 5 o'clock; then we drove in a car to Amsterdam, because it suited us best by car, spent the night in a hotel and had dinner there, a beer in the bar and then to bed. Next a strategic meeting in Holland; this lasted till five o'clock, after which to the airport in Amsterdam and then by plane to London, then to a hotel, a meeting with a German colleague at a hotel, dinner with this colleague, you know, a beer in the bar and up to the hotel. The next day a meeting with an American and the German and my European executive/boss, who in the meantime had been to Germany, and then in the evening back home.
>
> (Lassen 2006: 306)

Similar descriptions of 'life in corridors' (Lassen 2006: 306; see also Lassen (Chapter 9) in this volume) are found in other studies on the mobility praxis of knowledge workers (see Kesselring 2006b; Vogl 2007). Life in transit makes actual geographical location irrelevant. Direct contact with the social and physical environment is not factored into such mobility.

There seem to be equivalents in the realm of air travel to the lifestyles whose structural parameters were analyzed and critically examined by Schneider *et al.* (2002). Long-distance relationships between London and New York or Tokyo and Paris are no longer unusual. So-called 'NY-Londoners' (Doyle and Nathan 2001: 17) have an apartment in New York but spend most of their time in London (or vice versa). The 'everyday cosmopolitization' (Beck 2004) of modern life provided by air travel makes love and friendship spanning great distances no longer especially exotic or romantic. To be sure, it may not be a mass phenomenon, but living in and with mobility is becoming unspectacular and is no longer seen as a deviation from the norm (Urry 2007; Bauman 2005; Frändberg and Vilhelmson 2003). The spatial expansion of social networks will increase with the expansion of the provision of safe, reliable air travel by low-cost carriers (Groß and Schröder 2007). Likewise, cheap air travel enables migrants to maintain their family and social networks in their home country while spending their working lives someplace else. Just as the automobile has been an enabler of individualization, the aeroplane has become a tool enabling people to overcome space and distance and maintain original relationships (Gottdiener 2001, Doyle and Nathan 2001).

It is no accident that companies choose locations with quick access to airports (Pagnia 1992). Reaction times must be fast, and key employees must be able to be in the air quickly. McKinsey has an office in Munich airport. The idea of the 'company with its own airport', once used by a German airport as a promotion slogan, is persuasive enough to define structures and is already influential in urban and regional planning (Hartwig 2000; Graham and Marvin 2001).

Again, the importance of air travel should not be measured in terms of sheer passenger miles. Compared with the automobile, air travel plays a seemingly marginal role quantitatively. But the rates of increase are more than remarkable and point to a development that will not be without profound effect for the social matrix of modern societies and their ecological situation. At present the aeroplane is used by about four million people a day. The air space above the United States is populated by about 300,000 people at all times of the day and night, the equivalent of a medium-sized city in the air. About 1.6 billion flights are made yearly (Urry 2003: 154; Fuller and Harley 2005).

The major European airports – London Heathrow, Frankfurt, Paris Charles de Gaulle, Madrid Barajas and Amsterdam Schiphol – have had yearly passenger traffic increases of 7 to 8 per cent for years, doubling over thirteen years. In 1989 these airports were used by 116 million travelers, in 2003 by 234 million.[2] Frankfurt Airport moved around 30 million people in 1992, and 54 million in 2006. Airports in the second and third rank such as Munich, Copenhagen or Zurich are likewise registering steady increases (see O'Connor 2003). Twelve million people flew into or out of Munich in 1992; the 30 million mark was reached in 2006. Two years after the inauguration of a second runway in Munich in 2003, discussion began on the necessity of a third.

A functionally highly differentiated system of airports has developed, with varying degrees of importance (see Rodrigue *et al.* 2005). The business centers of the world have equally busy airports: '[K]ey cities are (re)produced by what flows through them rather than by what is fixed within them' (Derudder *et al.* 2005). The networked society is not found in virtual space alone, as Castells' (1996) argument sometimes supposes; it is a thoroughly material phenomenon too. Data and information flows connect with the physical flows of people and goods, whereby digitalization phenomona and materialization phenomena can hardly be analyzed and understood separately (see Lübbe 1995; Graham and Marvin 1996; Hanley 2004; Kitchin and Dodge (Chapter 5), in this volume).[3] Spatial mobility takes place within the parameters of 'time-space compression' (Harvey 1989) and proceeds hand-in-glove with the development and deployment of highly complex logistics and transport systems. The acceleration of social, economic and political processes brings about an ever more tightly knit network of relations between individuals and localities constituting a society based on a complex matrix of socio-material networks (Urry 2007, Kesselring 2006b) that connect with and stabilize one another.

Peter J. Taylor's world city network theory (Taylor 2004) deals with the connection between globalization and the formation of the networked society. In their discussion of the positioning of American cities on the global map, Derudder *et al.* (2005) cite five reasons why traffic flows in the worldwide network of airline routes provide the central data source for a geography of globalization. First, airline route maps are one of the few available indices for transnational traffic flows and interurban connectivities. Second, airline route networks and the supporting infrastructure are the most tangible manifestations of interaction between world cities. Third, demand for direct face-to-face contacts and meetings remains high, notwithstanding the telecommunications revolution. Fourth, air travel is the preferred means of transport for the transnational business elite, for tourists, migrants, and for high-value-added goods. And finally, airline connections are the central component in the international competition among cities and for a place in the sun among world cities.

There have been several attempts to describe the world city network in part or in full on the basis of numerical data provided by worldwide air traffic (see Keeling 1995; Smith and Timberlake 1995; Cattan 1995; O'Connor 1995; 2003). Witlox and Derudder have published the most solidly data-based research to date. Unlike others in the field, they measure the arrivals and departures at various airports and not merely flights between cities (see Derudder *et al.* (Chapter 4), in this volume). Their data show how frequently a city is the destination or the starting point of a flight. Other databases measure only movements between cities, saying nothing as to the destinations of travelers since every change of plane is registered as a separate flight. Derudder *et al.* make the axes of globalization clear. The most important axis is between London and New York. It is the only really relevant intercontinental connection; all the other major axes – those between New York and Los Angeles and between London and Paris or the Hong Kong-Singapore-Bangkok triangle – are intraregional. The maps in Derudder *et al.* in this volume show a decidedly Western bias. But new airlines are entering the market, and new airports are being built in the Middle East and Asia, so shifts in the configuration of globalization may be in the offing.

A number of basic conclusions about mobility and globalization may be drawn from this research. The activity radius of individuals, corporations and whole societies has expanded as a consequence of transnational network formation. The connectivity of the centers of business activity has grown with increases in the number of direct flights between global business cities. But at the same time we note the exclusive structure of global networked society, something which can also be seen in the geography of the Internet (see Castells 2001; Zook 2005). The 'Market Empire'[4] has appropriated the entire globe as its radius of activity, whereby geographical and digital spaces are not discrete entities but closely coupled spaces. This initially business-driven development has taken place hand-in-glove with profound cultural and social structural change. The globalization and cosmopolitization of societies are based on the existence of major international infrastructure systems such as the worldwide airline industry and the Internet, as well as transnational and trans-regional surface transport networks (cf. Hajer 1999; Jensen and Richardson 2004). Together they make up the mobility potential, the motility, of the mobile risk society in the age of second modernity (Canzler *et al.* 2008).

From monomodal air traffic terminals to global players: airports' great leap forward

Structural changes in mobility show a close correlation between air travel and the development of a globalized world. Brenner (1997: 12) points out that we may speak of globalization only if 'expansion, acceleration and other changes in capital accumulation demand the availability of large-scale territorial infrastructures like railways, superhighways, harbors, waterways, airports and state regulatory administrations that facilitate the ever quicker circulation of capital.' If we look at the worldwide expansion of the airport network, then this moment has arrived.

Airports are the backbone of global risk society. The modernization of the modern has reached a point where the framework of the nation-state has been left behind and stable transnational structures have taken its place (Jensen and Richardson 2004; Beck 2004). Globalized social, cultural, economic and political interaction has become everyday experience and has irreversibly changed the social morphology of cities and societies.

> The availability of relatively immobile transport, communication and regulatory-institutional infrastructures, i.e., of a 'second nature' consisting of socially generated configurations of territorial organization, makes this accelerated circulation of goods [as well as individuals and ideas; S.K.] in space possible.
>
> (Brenner 1997: 10)

Airports are the essential item in this 'second nature' in the process of the globalization of society and the economy. This is manifested in their appearance, organization and social significance. John F. Kennedy International Airport in New York and the airports at Barcelona, Madrid Barajas or Frankfurt can hardly be compared with their predecessors in the 1950s and 1960s. The transfer point airport has undergone a profound structural transformation:

> In the 1960s, the airport was considered an 'air train station'. In the 1970s it became an interface between air and rail traffic as well as a shopping center, in the 1980s it stressed its role as a business center and in the 1990s leisure time and entertainment have been prioritized.
>
> (Manfred Schölch, cited by Schamp 2002: 139ff.)

The economist David Jarach speaks of a first and second quantum leap in the transport and urbanizing functions of international airports. Formerly strictly monomodal and monofunctional traffic terminals, airports have been reinvented as multimodal hubs comparable with a 'multi-point, multi-service, marketing-driven firm' in a global marketplace (Jarach 2001: 119). Jarach dates the first such transformation in the 1970s, when airport authorities decided to leave their 'splendid isolation' behind and adopted the 'multimodal hub approach' (Jarach 2001: 121). This approach saw passengers and freight customers as demanding smooth transfers from air to ground transport, from aeroplane to rail, road and water transport. The airport, heretofore a peripheral transport-logistical facilitator for air travel, a non-place and ordinary transfer point, became the central instrument of national and transnational economic flows.

In the meantime, airports have undertaken a second quantum leap. Seamless mobility, the smooth performance of the complex processes and work steps involved in getting aeroplanes into the air and onto the ground safely and on time, is no longer so important. Airport operators such as the British BAA or Amsterdam Schiphol have their major revenues in 'non-aviation activities' (see Schamp 2002: 141). Jarach (2001) names five such non-aviation activities: commercial offerings (shopping of all kinds, from supermarket and duty free to fashion boutique); tourist offerings (hotels, gaming in Amsterdam, disco dancing at London Heathrow, Buddhist meditation in Lyon, plane spotter platforms, animated airport tours, family entertainment, restaurants, concerts, theater, etc.); business services offerings (convention centers, seminar and conference rooms, VIP lounges, etc.); logistics offerings (car rental, air freight, etc.); and knowledge consulting (airport operation expertise, airport construction expertise, etc.).

These examples show how the configuration of international airports has changed and has also become an object of interest for many more actors than just transport authorities and the transport industry. The publicly operated transfer point is disappearing (at least among international airports) and has been supplanted by the 'commercial airport' whose operators are entrepreneurs in a global business. Privatizations in the air transport sector will 'inevitably lead to cross-border airport ownership and the creation of multinational airport companies' (Doganis 2001, cited by Schamp 2002: 143). This describes the trend in air transport and airport development. Privatization of airlines (Burghouwt and Hakfoort 2002) is often followed by the partial or complete privatization of airports, as in the case of Britain's BAA (Francis and Humphreys 2001). Airports such as Frankfurt, Munich, Amsterdam, Heathrow or Manchester are run according to good business principles by managers who see themselves not as public officials but as profit-oriented businesspeople (Schamp 2002: 139).

This tendency is illustrated by the development of Frankfurt International Airport into a global player in the airport industry (for a detailed analysis see Schamp (2002) and Kesselring (2007)). Among the global hubs in Germany, the Fraport Corporation is the only publicly traded airport company. Going public enabled the company to realize two objectives: to shed all semblance of state ownership and tap into the capital markets. Fraport is a leading exponent of the commercial airport concept, not only in Germany. Together with the Amsterdam Schiphol Group, Fraport stands for a radical program in the competition for world market share in the air transport business. Fraport no longer sees itself simply as an institution that provides transport infrastructure; rather, it is a network of diverse firms and services operating in the context of the air travel sector. To be sure, Fraport performs all of the standard operations of an airport (ground servicing, check-in, baggage hand-ling, general aviation, etc.) but it also engages in all of those business activities characteristic of the second quantum leap.

Airport politics: the defining factor on the territorial level

It is not the case that there has been steady, uncontested progress towards global airport expansion and internationalization. The history of airport projects shows how complex and controversial the conflicts are today and have been since the 1960s and 1970s (Rucht 1984; Apter and Sawa 1984; Sack 2001; Troost 2003; Geis 2005). Guillaume Faburel's studies on controversies over airport interests in the United States and Europe (Faburel 2001; 2003) confirm this assessment and show how globalization and increasing demand for air travel have radicalized the issue. The theoretically relevant conclusion is that globalization processes do not proceed in linear fashion but rather are constantly challenged by opposing interests and can take unpredictable courses when they reach the territorial level. The spaces of globalization and the spaces of territorialization do not coexist in a simple relationship with one another; they are two sides of the same coin and represent different logics of decision-making and practical action in the process of the global restructuring of cities and regions.

> [P]rocesses of deterritorialization are not delinked from territoriality; indeed their very existence presupposes the production and continual reproduction of fixed socio-territorial infrastructures . . . within, upon, and through which global flows can circulate. Thus the apparent deterri-torialization of social relations on a global scale hinges intrinsically upon their reterritorialization within relatively fixed and immobile sociospatial configurations at a variety of interlocking subglobal scales.
>
> (Brenner 2004: 56)

What Brenner is saying is that globalization works itself out on concrete objects that seem to be relatively stationary and immobile. Airports are paradigmatic for this almost ontological dialectic of 'fixity and motion' that defines the specific spatiality of the global society (Harvey 1989; Jessop 2006; Brenner 1998). On the one hand, airports are interfaces with global space; they stabilize the cosmopolitan mobility potential of the mobile risk society by providing the logistic infrastructure for the acceleration and global coordination of organizational processes in business and society. But, on the other hand, airports are territorial and thus bound by the social, cultural, economic and political norms of their location. They cannot develop independently – hence, as Faburel (2003) notes on the basis of empirical studies in the United States and Europe, the importance of the neighboring local level. Regional forces often shape the planning stages of airport projects, especially so when expanding capacity is the issue. The resistance of neighboring residents can influence such operating parameters as take-off and landing directions, night operations, etc. Faburel lists a number of cases – Boston, Chicago, Los Angeles, San Francisco, Amsterdam, Paris – where local opposition led to the redefinition of larger and smaller projects. The local level, as other examples show (Deckha 2003; Flyvbjerg *et al.* 2003; Jensen and Richardson 2004), is anything but powerless and can successfully intervene and redefine projects in which global interests are involved:

> [B]y the values and legitimacies they carry, by the coalitions between elected officials in local communities who structure their action, these territories more and more effectively hinder the operators or proprietors of projects, enough to sometimes even redefine some of the political intentions of airport projects and management.
>
> (Faburel 2003, 1)

Occasionally expansion plans can bring about a radicalization in the political strategies of territorial power groups. Faburel (2003) describes a general trend away from the NIMBY ('not in my backyard') attitude towards the BANANA principle ('build absolutely nothing anywhere near anybody'). Local interests are becoming less important in political conflict situations involving the mobility that defines the mobile risk society. Instead, opponents are bringing to bear the values of life politics, which Giddens (1996) describes as the principles of radical democracy. Controversies surrounding airport projects articulate more and more the general interest of the informed citizenry in sustainable mobility policy. 'Sustainable aviation' (Thomas *et al.* 2003) is the issue, i.e. long-term, people-friendly and environmentally conscious management of global mobility flows. This can have structuring influence, for example when the 'geography of aeroplane noise' (Faburel 2001) manifests itself independently of the spatial geography surrounding an airport (see Hartwig 2000). The structuring influence of global mobility is especially apparent in discourse on its negative side effects. Controversies over airport noise, environmental pollution, land use, etc. affect the configuration of political, economic and societal networks in the territorial context (see Geis 2005). Thus global demand for airport capacity leads to the socio-material restructuring both of affected (sub)urban spaces and of the relationships between territorial political power centers and their networks.

According to Faburel (2003) the process of accommodating global interests (i.e. demand for air travel, the business objectives of airport operators, and the desire of cities and regions for positive global positioning) takes place in a matrix occupied by airport operators, local residents and local political authorities. Faburel analyzes the triangle of power and interaction provided by air transport operators, residents and local authorities (Figure 2.1). The first represent the interests of airports and the airline industry and thus, apart

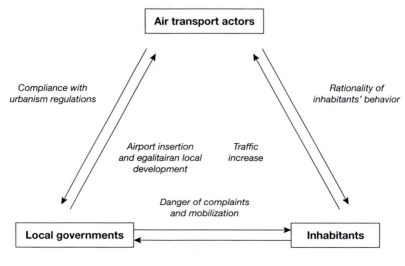

Figure 2.1 Scheme of contradictory positions between actors.
Source: Faburel 2003

from their own interests, those of airlines and of businesses that cater to the airport, airlines, airline passengers, etc. Airport operators especially play a strategically crucial role in negotiative processes with neighboring residents, government institutions and political power groups. In conflict situations, airport operators very often represent the whole aero-industrial complex, the *pôle aérien* as Faburel (2003) puts it.

Conclusion – politics in mobile risk society

The discussion to this point raises several questions relevant to the sociological analysis of political processes in the mobile risk society (Beck *et al.* 1999; Kesselring 2008). Two conclusions are especially important: First, in the future, modern institutions will be heavily occupied with the negative ecological effects of air transport, not least as it can be expected to gain in importance for modern lifestyles, cultures and economies. Lübbe (1995) has shown that the concentration of communication processes leads to the intensification of travel activities and is not a substitute for actual physical transport. The cosmopolitization of social and virtual networks goes along with an increase in long-distance travel activities, i.e. air travel (Frändberg and Vilhelmson 2003). Virtualization processes notwithstanding, physical proximity continues to be decisive for successful communication and stable interaction (see Urry 2002), which is why actual physical transport will continue to grow in modern societies. Concentration processes in space and time and global network formation are progressive; the world city network continues to grow, and the concentration and intensification of global travel networks will also continue apace (see Harvey 1989; Castells 1996; Taylor 2004). Globalized forms of work and production, transnationalization and intertwining networks of corporations and industries, and the unbounding connective effects of IT and telecommunication will enhance the relevance of cross-border relationships and cross-border interaction (see Tomlinson 2003; Thierstein *et al.* 2006). Hence, second, we may be sure that airport operators and the entire aero-industrial complex (*pôle aérien*) will continue to press their expansion and internationalization strategies. The pressure on cities and regions to increase airport capacities will not let up. Fraport predicts demand for up to 700,000 take-offs and landings per year by the year 2020, an increase of more than 200,000 compared with today's figure (Fraport 2006: 44). This demand can only be met by building a fourth runway, now in planning. But negotiating systems such as the discursive practice of mediation procedures now in place have proven to be highly ambivalent conflict cultures that neutralize linear political leadership (Giegel 1998). It seems that government, after the 'negative political experiences it had in the past[,] . . . would rather forgo political leadership voluntarily' (Geis 2005: 130). The social construction of global mobility potentials proves to be a complex global-local process in which participants on all levels compete for defining influence (Sack 2001).

The mediation procedure of the years 1999–2002 concerned with airport expansion in Frankfurt is an instructive example for the complexity of contemporary deliberative mobility politics. It was an attempt to resolve the conflicts between the above-mentioned power triangle of airport operators, residents and local authorities. It was the most extensive mediation ever attempted in Germany and has been judged highly critically as to its outcome. Geis (2005) sees the process has having pacified the conflict and de-emotionalized the issues. But the process was also mentioned as having 'claimed a consensus that did not exist' (Troost 2003: 179). According to Troost, the procedure did not succeed in including all relevant parties and hence is diminished in legitimacy and authority. Since certain environmental groups and citizen action groups refused to participate in the mediation process, their members represent an unpredictable and highly ambivalent protest potential when the building phase begins. After the effective decision in December 2007 to construct a new runway it will show the quality and the sustainability of deliberative politics in airport conflicts.

Many analyses (see Sack 2001; Troost 2003; Geis 2005) posit a fundamental indecisiveness on the part of German politicians, as in other countries, as to the proper course to take in the question of the inexorable development of global mobility potentials. The negative consequences of increased air travel have been precisely analyzed (Thomas *et al.* 2003), and it is common knowledge that the burden of globalization lies on local contexts. The German government's airport white paper, published in 2000 (BMVBW 2000), has brought about no significant future-oriented changes in airport policy, but it accurately notes that airlines increasingly demand airport capacity at competitive prices. If this demand cannot be met at certain locations or for certain regions, traffic flows – especially plane changes at the major transfer points – will shift to other locations. For macroeconomic reasons, airport policy must thus see to it that Germany, with its multicenter airport system, in particular with the major transfer points at Frankfurt and Munich, maintains its attractiveness as a world business location through the provision of sufficient airport capacities. Only then will uninterrupted air connections to attractive locations be maintained. (BMVBW 2000: 30ff.)

The power of the airlines to set the agenda is accepted without question, and now it is the task of cities and local power structures to find solutions to the conflicts that arise. The consequences are 'interpretive offensives' (Geis 2005: 161) on the part of local groups to frame the problem definitions and to struggle for alternative problem solutions. The 'integrative transportation policy' often called for (Schöller 2006) is nowhere to be seen. Instead of a comprehensive concept, we encounter discourse practices and discourse coalitions trying to define paths of development and trying to find corridors of resolution.

By passing the pressure on to the territorial level, the national government intensifies the regional conflict potential, burdening local power structures with acquiescing to global interests or, equally onerous, refusing to do so. To the degree that pressure to expand airports increases, resistance is deployed against expansion. The *Wiesbadener Kurier*, a local German newspaper, reported on 23 May 2006 that 127,000 objections had been submitted during the construction plan approval process for the fourth Frankfurt airport runway. Such resistance is well within the democratically legitimated bounds of citizen participation. Studies on the politics of reterritorialization (Jessop 2001; Deckha 2003; Faburel 2003; 2001; Brenner 2004) show local groups as effective shapers of global mobility potential. Local players are not the helpless pawns of overwhelmingly powerful globalizing forces. They have proven themselves to be competent, power-conscious, strongly argumentative (micro- and sub-) political actors.

Clearly, hard thinking is necessary to arrive at a sustainable policy for the management of global mobility potentials. There are basically three approaches: first, technical solutions; second, avoidance strategies; and, finally, institutional initiatives aimed at the innovative restructuring of the issue. Technical improvements such as more efficient and quieter engines and noise prevention measures are of course necessary and to a certain extent helpful, but technical improvements are not enough. Avoidance strategies for the reduction of air traffic run counter to the interests of airport operators and airlines, but they are necessary to reduce the negative ecological effects of air travel, especially effects on the upper atmosphere and on climate change. Companies and public institutions must rethink their mobility management with an eye to reducing air travel. But this approach is likewise too little too late to solve the problem.

Efforts to modernize the relationships between the interested parties and to improve the quality of discourse on airport and air traffic issues are a more promising approach. If national governments continue to abstain in the formulation of airport policy, as in Germany, there is no alternative to a radical new concept of global-local policies on the management of airborne mobility. Without a comprehensive integrative concept, the local and regional levels have no choice but to seek new avenues leading to consensus at the interface of global and local interests. The Frankfurt mediation procedure is thus not so much an exercise in futility as an important precedent for 'glocal' politics (Berndt and Sack 2001). The question as to the role of global actors (airlines, interest groups such as the Board of Airline Representatives, ecological and social NGOs, etc.) is yet to be answered. To date, research has concentrated on the local level, with little or no penetration of the ways global players work to have their aims and interests expressed in locality-based airport policy.

Research on mobility politics in cities shows that the innovative potential inherent in the modernization of relationships between interested parties is considerable, but that it is not being fully exploited. Research to date on institutional innovation in regional and community-level transportation policy (Flämig *et al.* 2001; Kesselring *et al.* 2003) causes one to wonder whether the problems proceeding from the ecological and social consequences of mass (air) transport will bring about appropriate institutional responses. A German Federal Ministry of Research project on mobility in urban agglomerations has produced sobering results. Major research projects with strong emphasis on technology innovation such as Mobinet in Munich or WAYflow in the Frankfurt region[5] have not led to a mobility policy that ventures innovative solutions to transport problems (Kesselring 2001; Kesselring *et al.* 2003). To be sure, there have been institutional innovations such as Cooperative Transport Management or the so-called 'Inzell Circle' in Munich, which have provided important suggestions for changes in the make-up of political forums and have energized the deliberative process (Hajer and Kesselring 1999). But in the end, hard decisions by local authorities were simply transferred to a pre-political discourse and decision-making process. Complex decisions were presented to local authorities as cut and dried and needing only to be implemented (Vogl and Kesselring 2002).

On the other hand, pre-political processes are often more competent than institutionalized processes and closer to the complexity of mobility questions. Such deliberative practices have the potential to form new discourse coalitions that can integrate heretofore marginal positions into the political decision-making process and present them to the political and discursive mainstream (Hajer 1995; Richardson 1996; Jensen 2006). It is surely a positive development in the political culture when institutional business-as-usual gives way to issue-based discourse and new ideas enter the arena as legitimate possibilities. This can – as happened in Munich – even depolarize the debate and encourage the development of more rational discourse cultures (Healey 1993; Kesselring 2001).

One of the features of the mobile risk society (Beck *et al.* 2003; Kesselring 2008) is the creation of deliberative contexts and networks in which forms of knowledge that are outside the mainstream get a hearing (Hajer and Wagenaar 2003). The fact that the local level has such power in the struggle over global mobility potentials is an example of this. Any widely acceptable concept of politics in the mobile risk society must take this into consideration. Transport policy is made to order for 'life politics' and the articulation of positions and arguments not easily labeled 'left' or 'right'. Globalization has changed the topographies of the political (Hajer and Wagenaar 2003: 9) and created new arenas and forums in which territorial and global players maneuver for the upper hand. The open question that is posed in connection with airports as the global transfer points and that can only be answered empirically is: must we only expand the participatory features of representative democracy to include local participants in the process of finding consensus? Or must we progress and increase the transparency of the planning and decision-making processes to demystify the role of global players? At present we may assume that airport operators, who bear the main burden of political argument and negotiation, are only 'interscalar' and intermediate points along a line that begins and ends with bigger players in the global network society. For a contemporary sociology of mobilities, as well as for the politics of the mobile risk society, the challenge is to decode the complexities of the strategies global players use to influence the social and political construction of global mobility potentials such as airports.

Notes

1   Source: www.wikipedia.de.
2   Source: Bundesministerium für Verkehr (2006).
3   Dodge and Kitchin (2004) use the example of the connection between flying and IT background activities, e.g. booking and ticketing. They speak of 'code/space', of flying through IT-based or IT-generated space. Thrift (2004) posits a similar thesis, referring generally to the 'encoding' of geographic and social space, which he calls 'movement-space'.
4   I borrow the term from de Grazia (2006).
5   MOBINET was one of five big research and development projects in Germany financed by the German government. It took place in the Greater Munich region. In 2000 MOBINET received the first European Mobility Award in Paris. The €40 million R&D project was a major attempt to improve the regional organization of urban transport. WAYFLOW in the Frankfurt region was a similar project with comparable goals. It was also financed by the German government. Together with three other major projects in Dresden, Stuttgart and Cologne, MOBINET and WAYFLOW represented pioneering attempts to take new paths in technology and transport policy.

Bibliography

Adey, P. (2006) 'If Mobility is Everything, Then It is Nothing: Towards a Relational Politics of (Im)mobilities', *Mobilities*, 1(1): 75–95.
Apter, D. E. and Sawa, N. (1984) *Against the state: politics and social protest in Japan*, Cambridge, Massachusetts: Harvard University Press.

Asendorf, C. (1997) *Super Constellation – Flugzeug und Raumrevolution. Die Wirkung der Luftfahrt auf Kunst und Kultur der Moderne*, Vienna/New York: Springer.

Bauman, Z. (2005) *Liquid Life*, Cambridge: Polity Press.

Beck, U. (1992) *Risk Society*, London: Sage.

—— (2000) *What is globalization?*, Cambridge: Polity Press.

—— (2004) *Der kosmopolitische Blick oder: Krieg ist Frieden*. Frankfurt/Main: Suhrkamp.

—— (2008) 'Mobility and the Cosmopolitan Perspective', in W. Canzler, V. Kaufmann and S. Kesselring (eds.) *Tracing Mobilities*, Aldershot: Ashgate.

—— Hajer, M. and Kesselring, S. (eds.) (1999) *Der unscharfe Ort der Politik. Empirische Fallstudien zur Theorie der reflexiven Modernisierung*, Opladen: Leske & Budrich.

—— Bonss, W. and Lau, C. (2003) 'The Theory of Reflexive Modernization: Problematic, Hypotheses and Research Programme', *Theory, Culture & Society*, 20 (2):1–34.

Berndt, M. and Sack, D. (2001) *Glocal governance? Voraussetzungen und Formen demokratischer Beteiligung im Zeichen der Globalisierung*. Wiesbaden: Westdeutscher Verlag.

BMVBW (Bundesministerium für Verkehr, Bau- und Wohnungswesen) (2000) *Flughafenkonzept der Bundesregierung vom 30. August 2000*. Bonn.

Brenner, N. (1997) 'Globalisierung und Reterritorialisierung: Städte, Staaten und die Politik der räumlichen Redimensionierung im heutigen Europa', *WeltTrends*, 17: 7–30.

—— (1998) 'Between Fixity and Motion: Accumulation, Territorial Organization and the Historical Geography of Spatial Scales', *Environment and Planning D: Society and Space*, 16: 459–81.

—— (2004): *New State Spaces. Urban Governance and the Rescaling of Statehood*, Oxford: Oxford University Press.

Brueckner, J. K. (2003) 'Airline Traffic and Urban Economic Development', *Urban Studies*, 40 (8):1455–69.

Bundesministerium für Verkehr (2006) *Verkehr in Zahlen 2005–2006*. Bonn: Deutsches Institut für Wirtschaftsforschung.

Burghouwt, G. and Hakfoort, J. A. (2002) 'The Geography of Deregulation in the European Aviation Market', *Tijdschrift voor Economische en Sociale Geografie* 93 (1): 100–6.

Canzler, W., Kaufmann, V. and Kesselring, S. (eds.) (2008) *Tracing Mobilities. Towards a Cosmopolitan Perspective*. Aldershot: Ashgate.

Castells, M. (1996) *The Rise of the Network Society*, Oxford: Blackwell Publishers.

—— (2001) *The Internet Galaxy: Reflections on the Internet, Business, and Society*, Oxford: University Press.

Cattan, N. (1995) 'Attractivity and Internationalization of Major European Cities: The Example of Air Traffic', *Urban Studies*, 32(2): 303–12.

Deckha, N. (2003) 'Insurgent Urbanism in a Railway Quarter: Scalar Citizenship at King's Cross, London', *ACME, An International E-Journal for Critical Geographies*, 1(2): 33–56.

De Grazia, V. (2006) *Irresistible Empire: America's Advance Through Twentieth-Century Europe*, Cambridge, MA: Harvard University Press.

Derudder, B. and Witlox, F. (2005) 'An Appraisal of the Use of Airline Data in Assessing the World City Network: A Research Note on Data', *Urban Studies*, 42(13): 2371–88.

—— and Taylor, P. T. (2005) 'United States Cities in the World City Network: Comparing Their Positions Using Global Origins and Destinations of Airline Passengers', *GaWC Research Bulletins*, 173.

Dicken, P. (2003) *Global Shift. Reshaping the Global Economic Map in the 21st Century*, New York/London: The Guildford Press.

Dodge, M. and Kitchin, R. (2004) 'Flying Through Code/Space: The Real Virtuality of Air Travel', *Environment and Planning*, 36(2): 195–211.

Doganis, R. (2001) *The airline business*, London: Routledge.

Doyle, J. and Nathan, M. (2001) *Wherever Next? Work in a Mobile World*, London: The Industrial Society.

Faburel, G. (2001) *Le bruit des avions. Évaluation du coût social: Entre aéroport et territoires*, Paris: Presses École Nationale Ponts Chaussées.

—— (2003) 'Lorsque les territoires locaux entrent dans l'arène publique: Retour d'expériences en matière de conflits aéroportuaires', *Espaces et Sociétés*, 115: 123–46.

Flämig, H., Bratzel, S., Arndt, W. H. and Hesse, M. (2001) *Politikstrategien im Handlungsfeld Mobilität. Politikanalyse von lokalen, regionalen und betrieblichen Fallbeispielen und Beurteilungen der Praxis im Handlungsfeld Mobiltät*. IÖW-Schriftenreihe Nr. 156/01. Berlin.

Flyvbjerg, B., Bruzelius, W. and Rothengatter, N. (2003) *Megaprojects and Risks. An Anatomy of Ambition*. Cambridge: University Press.

Francis, G. and Humphreys, I. (2001) 'Airport Regulation: Reflecting on the Lessons from BAA plc', *Public Money & Management*, 21(1):49–52.

Frändberg, L. and Vilhelmson, B. (2003) 'Personal Mobility: A Corporeal Dimension of Transnationalisation. The Case of Long-distance Travel from Sweden', *Environment and Planning*, 35(10): 1751–68.

Fraport (2006) *Frankfurt Airport – Luftverkehrsstatistik 2005*. Frankfurt/Main.

Fuller, G. and Harley, R. (2005) *Aviopolis. A Book About Airports*, London: Black Dog Publishing.

Geis, A. (2005) *Regieren mit Mediation. Das Beteiligungsverfahren zur zukünftigen Entwicklung des Frankfurter Flughafens*. Wiesbaden: VS Verlag für Sozialwissenschaften.

Gerstenberger, H. and Welke, U. (2002) *Seefahrt im Zeichen der Globalisierung*, Münster: Westfälisches Dampfboot.

Giddens, A. (1996) *Beyond Left and Right: The Future of Radical Politics*, Stanford, California: Stanford University Press.

Giegel, H. J. (1998) *Konflikt in modernen Gesellschaften*, Frankfurt/Main: Suhrkamp.

Gottdiener, M. (2001) *Life in the Air. Surviving the New Culture of Travel*, Lanham, Maryland: Rowman and Littlefield.

Graham, S. and Marvin, S. (1996) *Telecommunications and the City. Electronic Spaces, Urban Places*, London, New York: Routledge.

—— and —— (2001) *Splintering Urbanism. Networked Infrastructures, Technological Mobilities and the Urban Condition*, London, New York: Routledge.

Groß, S. and Schröder, A. (2007) *Handbook of Low Cost Airlines Strategies, Business Processes and Market Environment*, Berlin: Erich Schmidt Verlag.

—— and Wagenaar, H. (2003) *Deliberative Policy Analysis. Understanding Governance in the Network Society*, Cambridge: University Press.

—— and Kesselring, S. (1999) 'Democracy in the Risk Society? Learning from the New Politics of Mobility in Munich', *Environmental Politics*, 8(3): 1–23.

Hajer, M. (1995) *The Politics of Environmental Discourse. Ecological Modernization and the Policy Process*, Oxford: Oxford University Press.

Hajer, M. (1999): 'Zero-Friction SocietyUrban', *Design Quarterly*, 71: 29–34.

Hanley, R. (2004) *Moving People, Goods, and Information in the 21st Century: The Cutting-edge Infrastructures of Networked Cities*, London, New York: Routledge.

Hannam, K., Sheller, M. and Urry, J. (2006) 'Mobilities, Immobilities and Moorings. Editorial', *Mobilities*, 1(1): 1–22.

Hartwig, N. (2000) *Neue urbane Knoten am Stadtrand? Die Einbindung von Flughäfen in die Zwischenstadt: Frankfurt/Main – Hannover – Leipzig/Halle – München*, Berlin: VWF Verlag für Wissenschaft und Forschung.

Harvey, D. (1989) *The Condition of Postmodernity. An Enquiry into the Origins of Cultural Change*, Cambridge/Oxford: Blackwell.

Healey, P. (1993) 'Planning Through Debate: The Communicative Turn in Planning Theory', in F. Fischer and J. Forester (eds.) *The Argumentative Turn in Policy Analysis*, Durham: Duke University Press.

Jarach, D. (2001) 'The Evolution of Airport Management Practices: Towards a Multi-point, Multi-service, Marketing-Driven Firm', *Journal of Air Transport Management*, 7: 119–25.

Jensen, A. (2006) 'Governing with Rationalities of Mobility', Ph.D. dissertation, University of Roskilde, Denmark.

Jensen, O. B. and Richardson, T. (2004) *Making European Space. Mobility, Power and Territorial Identity*, London, New York: Routledge.

Jessop, B. (2001) 'Institutional (Re)turns and the Strategic-Relational Approach', *Environment and Planning A*, 33(7): 1213–35.

—— (2006) 'Spatial Fixes, Temporal Fixes, and Spatio-temporal Fixes', in N. Castree and D. Gregory (eds.) *David Harvey: A Critical Reader*, Oxford: Blackwell Publishing.

Keeling, D. J. (1995) 'Transport and the World City Paradigm' in P. L. Knox and P. J. Taylor (eds.) *World Cities in a World System*, Cambridge: University Press.

Kesselring, S. (2001) *Mobile Politik. Ein soziologischer Blick auf Verkehrspolitik in München*, Berlin: edition sigma.

—— (2006a) 'Global Transfer Points. International Airports and the Future of Cities and Regions', paper presented at Air Time-Spaces. New Methods for Researching Mobilities, 29–30 September 2006, Lancaster University.

—— (2006b) 'Pioneering Mobilities. New Patterns of Movement and Motility in a Mobile World', *Environment and Planning, Special Issue on Mobilities and Materialities*: 269–79.

—— (2007) 'Globaler Verkehr – Flugverkehr', in O. Schöller, W. Canzler and A. Knie (eds.) *Handbuch Verkehrspolitik*, Wiesbaden: VS Verlag.

—— (2008) 'The Mobile Risk Society. Mobility and Ambivalence in the Second Modernity', in W. Canzler, V. Kaufmann and S. Kesselring *Tracing Mobilities*. Aldershot, Burlington: Ashgate.

——, Moritz, E. F., Petzel, W. and Vogl, G. (2003) *Kooperative Mobilitätspolitik. Theoretische, empirische und praktische Perspektiven am Beispiel München und Frankfurt, Rhein/Main.* München: IMU.

Knox, P. L. and Taylor, P. J. (eds.) (1995) *World Cities in a World-System.* Cambridge, New York: Cambridge University Press.

Lassen, C. (2006) 'Aeromobility and Work', *Environment. and Planning*, 38(2): 301–12.

Lefebvre, H. (2000) *The Production of Space.* Oxford, Cambridge: Blackwell Publishing.

Lübbe, H. (1995) 'Mobilität und Kommunikation in der zivilisatorischen Evolution', in Spektrum der Wissenschaft. Dossier 2, S. 112–19.

O'Connor, K. (1995) 'Airport Development in Southeast Asia', *Journal of Transport Geography*, 3(4): 269–79.

—— (2003) 'Global Air Travel: Toward Concentration or Dispersal?', *Journal of Transport Geography*, 11: 83–92.

Oswalt, P. (2004) *Schrumpfende Städte. Vol. 1: Internationale Untersuchung. Kulturstiftung des Bundes.* Ostfildern: Hatje Cantz.

Pagnia, A. (1992) *Die Bedeutung von Verkehrsflughäfen für Unternehmungen: eine exemplarische Untersuchung der Flughäfen Düsseldorf und Köln/Bonn für Nordrhein-Westfalen.* Frankfurt/Main, Berlin, Bern, Brussels, New York, Oxford, Vienna: Lang.

Richardson, T. (1996) 'Foucauldian Discourse: Power and Truth in Urban and Regional Policy Making', *European Planning Studies*, 4(3): 279–92.

Robertson, R. (1992) *Globalization. Social Theory and Global Culture,* London: Sage.

Rodrigue, J. P., Comtois, C. and Slack, B. (2005) *The Geography of Transport Systems,* London, New York: Routledge.

Rucht, D. (ed.) (1984) *Flughafenprojekte als Politikum: die Konflikte in Stuttgart, München und Frankfurt,* Frankfurt, Main, New York: Campus Verlag.

Sack, D. (2001) 'Jobs, Lärm und Mediation. Zur demokratischen Partizipation bei glokalen Großprojekten' in M. Berndt and D. Sack (eds.) *Glocal Governance,* Opladen: Westdeutscher Verlag.

Sassen, S. (1991) *The Global City,* New York, London, Tokyo: Princeton University Press.

Schamp, E. W. (2002) 'From a Transport Node to a Global Player: The Changing Character of the Frankurt Airport', in D. Felsenstein, E. W. Shachar and A. Schamp (eds.) *Emerging Nodes in the Global Economy: Frankfurt and Tel Aviv Compared,* Dordrecht, Boston, London: Kluwer.

Schneider, N. F., Limmer, R. and Ruckdeschel, K. (2002) *Mobil, flexibel, gebunden. Familie und Beruf in der mobilen Gesellschaft,* Frankfurt/Main: Campus Verlag.

Schöller, O. (2006) *Mobilität im Wettbewerb. Möglichkeiten und Grenzen einer integrativen Verkehrspolitik im Kontext deregulierter Verkehrsmärkte,* Düsseldorf: Hans-Böckler-Stiftung.

Smith, D. A. and Timberlake, M. (1995) 'Conceptualizing and Mapping the Structure of the World System's City System', *Urban Studies*, 32(2): 287–302.

Swyngedouw, E. (1997) 'Neither Global nor Local: "Glocalization" and the Politics of Scale', in K. R. Cox (ed.) *Spaces of Globalization. Reasserting the Power of the Local,* New York: Guilford Press.

Taylor, P. J. (2004) *World City Network. A Global Urban Analysis*, London, New York: Routledge.

Thierstein, A., Kruse, Ch., Glanzmann, L., Gabi, S. and Grillon, N. (2006) *Raumentwicklung im Verborgenen: die Entwicklung der Metropolregion Nordschweiz*, Zürich: Verl. Neue Zürcher Zeitung (NZZ Libro).

Thomas, C., Upham, P., Maughan, J. and Raper, D. (2003) *Towards Sustainable Aviation*, Sterling: Stylus Pub Llc.

Thrift, N. (2004) 'Movement-Space: The Changing Domain of Thinking Resulting from the Development of New Kinds of Spatial Awareness', *Economy & Society*, 33(4): 582–604.

Tomlinson, J. (2003) 'Culture, Modernity and Immediacy', in U. Beck, N. Sznaider and R. Winter (eds.) *Global America? The cultural consequences of globalisation*, Liverpool: Liverpool University Press.

Troost, H. J. (2003) 'Steuerung in der metropolitanen Region – clash or consensus?', in S. Buckel, R. M. Dackweiler and R. Noppe (eds.) *Formen und Felder politischer Intervention – Zur Relevanz von Staat und Steuerung. Festschrift für Josef Esser.* Münster: Westfälisches Dampfboot.

Urry, J. (2002) 'Mobility and Proximity', *Sociology*, 36(2): 255–74.

—— (2003) *Global Complexity*, Cambridge: Polity Press.

—— (2007) *Mobilities*, Cambridge: Polity Press.

Vogl, G. (2007) 'Selbstständige Medienschaffende in der Netzwerkgesellschaft. Zwischen innovativer Beweglichkeit und flexibler Anpassung', Ph.D. dissertation, Munich Technical University.

Vogl, G. and Kesselring, S. (2002): 'Reflexive Mobilitätsplanung. Soziologische Anmerkungen zum Leitprojekt MOBINET des Bundesforschungsministeriums', *RaumPlanung*, 103: 189–92.

Zook, M. A. (2005) *The Geography of the Internet Industry: Venture Capital, Dot-coms, and Local Knowledge*, Oxford: Blackwell Publishers.

Zorn, W. (1977) 'Verdichtung und Beschleunigung des Verkehrs als Beitrag zur Entwicklung der "modernen Welt" ', in R. Koselleck (ed.) *Studien zum Beginn der modernen Welt*, Stuttgart: Klett-Cotta.

Part II

The production of airspaces

3 > store > forward >

Architectures of a future tense

Gillian Fuller

If the acme of success in design is to 'achieve ubiquity, to become banal' then the airport has achieved what Bruce Mau would term 'design nirvana' (Mau and The Institute without Boundaries 2004: 3). Despite the spectacular engineering and monumental flashiness of many modern airports, the basic conceptual architecture for facilitating aeromobility has become pervasive. We walk–we stop–we sit–we walk–we stop. We progress through a set of procedures that facilitate mobility. This rhythm of stopping and going, waiting and then moving in climate-controlled, closely surveilled environments is familiar because, increasingly, we experience it in all traffic. This is the global space of flows. It stops a lot. And yet, beyond complaints about 'jams' and delays, we tend not to focus on how much 'stillness' is operationally required for a high-speed traversal through this planet's aeromobile environments. From the packing of clothes in fixed containers to strapping your belt – tight and low – stillness and all its requisite activities, technologies and behaviours are fundamental to the 'flow' architectures that organize the motion of the globalizing multitudes of today.

In *The Rise of the Network Society*, Vol. 1, Manuel Castells elaborates the term 'space of flows' as 'the material organization of time-sharing social practices that work through flows' (1998: 412). These practices synchronize land and air, flesh and code, pedestrian and passenger. Organized through information technology, the space of flows comprises 'repetitive, program-mable sequences of exchange and interactions between physically disjointed positions held by social actors' (1998: 412). In other words, the space of flows is also organized like information. If one considers an airport, almost everything about it is informational. Not only is the airport *saturated with* information in terms of signage, advertising, pricing labels, it is *managed through* real-time integrated information systems exemplified through the workings of air traffic control. But more significantly for this chapter, the airport is *organized like* information. Specifically, airports are situated within complex interactive systems that potentially connect any point in the system to another. In this ontology, informational concepts such as signal/noise, feedback, fluctuation and replication constitute a dynamic environment of constantly shifting flows. As Tiziana Terranova puts it: 'cultural processes are taking on the attributes of information – they are increasingly grasped and conceived in terms of their informational dynamics' (Terranova 2004: 7).

Packeted, tagged and routed through the networks of supermodern life (Augé 1995: 28–31), this chapter will explore just one aspect of the 'time-sharing social practices' of the space of flows: the time-sharing practice of waiting. This chapter will argue that such practices are predicated on a shared 'future' that pre-empts the present – a future that paradoxically stops us in our tracks. To look at how mobility is organized in the airport is to consider how movement – from the predictive and proscriptive logistics of global aviation to how a body creeps forward in a queue – is becoming informationalized. That is to say, movement is mobilized upon principles of 'probabilistic *containment* and *resolution* of the instability, uncertainty and virtuality of a process' (Terranova 2004: 24; my emphasis).

The information age produces, among other things, excessive information. Every moment of the day can be preserved on a cell-phone camera, every chat history can be archived, thus proliferating multiple pasts. Our futures are similarly prolific – a whole range of options of the things we could do are modelled, from our likely path through an airport to our potential book selections. How does one process, filter and navigate through the onslaught of options, locations and destinations now available? In a society in which control extends into new domains each day, perhaps all the predictive soft-wares and helpful algorithms will result in what Brian Holmes has called 'cybernetic governance' (2007), in which all desires and catastrophes are anticipated and pre-emptively resolved. The complex temporal scape we now inhabit seems dominated by anticipation and pre-emptive responses offering 'solutions' to the excesses of daily life that result in a future tense constantly looming in the immediate present.

Store and forward

The airport is a distribution architecture – it organizes the chaotic movement of bodies, planes, baggage and bits into sequenced flows through the protocols of store and forward. Stored in the departure lounge, forwarded onto the plane, and stored in the plane that forwards to the next sequence of hold and release. One waits in the departure lounge, at traffic lights, in the taxi queue, for the visa to be granted, and, for those of us from geographically remote places, counting the hours, 22, 21, 20 . . . hours, till we arrive. And then we wait again, in the immigration queue, in the taxi queue and at traffic lights. Life feels more like a constant holding pattern, but if one waits, if one is patient, one will eventually get 'there'. Incorporated by a destination we move into an informationalized milieu that is distributive and biotechnical, in which the now is suspended in forward planning and in which multiple metabolic and logis-tical times are meshed. Calibrated by the logic of store and forward, stillness and movement are synthesized into an operational unity that has become naturalized.

And yet, if a baggage system breaks down, if a plane is stuck on a runway, or if Heathrow goes on 'critical alert', the synchronization of events goes awry and the apparent transparency of the store and forward systems by which we move becomes visible and felt. Stuck, no longer distracted by forward movement into the next sequence (check-in, immigration, departure lounge, plane) we might experience another sense of time, we might become attentive to where we are rather than what lies ahead of us. We may be able to reassess the now, or at least wonder where it has gone.

On one level, one could say the airport is a meshwork of queues that interoperate at different modalities, tempos and dimensions. It processes series of waitings (for planes to board, for take-off clearance, for baggage to appear on the carousel, for a connecting flight, for taxis) each with its own material architectures, critical protocols and ambient affects. Store and forward is not a new technique. One might think of the Carthaginian prowess in making pots and sailing ships as an early win for store and forward systems – an ancient method of organizing movement that persists.

Meshing mobilities: temporal relations

'[N]ature as perceived always has a ragged edge' (Whitehead 2004: 50). In *The Concept of Nature*, Alfred Whitehead uses this metaphorical statement to describe the 'unexhaustive character' of knowledge, which for Whitehead is 'a stubborn fact' (1985: xiv). Whitehead's process philosophy, in which we know what we sense and in which 'relatedness' dominates over 'quality', speaks compellingly to the complex modulations inherent in networked environments. The airport is a metastable system, it is more than a building: it is an experience. The airport is so monumental and so integrated into multiple other systems we can only perceive it in fragments. We know there is more – things keeps poking through, seeping into and overlapping the seemingly neat boundaries that logistical architecture imposes on life to facilitate order. In the lines and clusters, the stops and starts of flow architectures that dominate modernity one sees the ragged edge of logistical life – a complex bioinformatic hybrid that is supported by multiple time-space relations that mesh technologies of motion, vision, navigation, informatics and risk management.

Consider the passenger's progress through the airport from landside to airside. The move from road to runway is fretted with a series of holding areas, in the forms of queues and lounges, during which the traveller transitions from local time and its accompanying routines to the more modelled and measured time of global airspace. As the passenger proceeds through check-in, immigration and security, he or she gradually emerges into the 'sterile' zone of airside shopping zones and departure lounges. During this process, the citizen subject is incorporeally transformed into a PAX- 'a generic passenger with no identifying marks' (Cresswell 2006: 238) – an abstract model of movement, subject to the exceptional rules of airspace. As many theorists on airports have noted, so-called 'real space' and code space hybridize at the airport

(Dodge and Kitchen 2004; Fuller and Harley 2004; Cresswell 2006). Thus the passenger not only undergoes the process of becoming a flesh avatar of their real self that resides in a database, but there is also a temporal aspect to this panoptic sorting. Despite the ongoing development of non-voluntary biometric systems that require no cooperation from 'the target', the technology is not sufficiently efficient. Therefore, we still need to stop in order to get sorted. We collude in our capture, because we learn soon enough that when we cooperate with the machine, it works better.[1]

Airports mesh both abstractly and materially, producing wildly hetero-genous spaces and mobilities. Thus to speak of an airport is not to speak of structure but to talk about a dynamic complexity of relations between multiple assemblages in which informational and social relations are in constant flux. One must go off one network (i.e. cell network) to go on another (the plane). One foregoes one paradigmatic set of relations (such as a right to privacy) for another (the 'privilege' of moving).

However, for the traveller, navigating such an ontologically unknowable meshwork is not the adventure of the senses that potentially it could be, rather it becomes a journey through statistical probabilities and informational redundancy. The airport may not be *knowable*, but it works to make itself *known* to the shifting components (planes, people, baggage) that comprise its form at any moment. Airports work on protocol and therefore necessarily reveal themselves in bits as the traveller gradually traverses a procedural landscape of check-in, verification and search sequences that accretively grant permission to fly. Only the wealthiest jump the queue.[2]

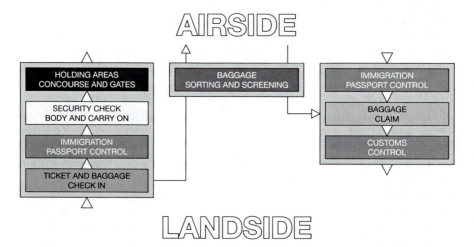

Figure 3.1 STORE FORWARD: Moving from airside to landside, the passenger is propelled through a series of holding areas and gates.

Source: Aviopolis Project

Anticipatory architecture

In a seminal sociological study on queuing and waiting, Barry Schwartz considered the waiting body as a locus of politics by exploring the distribution of time in social systems: 'the main proposition is the distribution of waiting time coincides with the distribution of power' (Schwartz 1975: 5). The airport inculcates many of the meanings attributed to waiting in Schwartz's study: a sign of corporate greed and aggression (when there are few servers); a type of disciplinary scaffold effect as we wait to be frisked like those ahead of us in the line; sometimes as a mark of deference to the procedure in air travel (as people obediently arrived three or even four hours early for flights immediately after September 11, 2001); or perhaps as a sign of value that we place on the convenience of air travel. We will lose hours now so we can be sped up later. Waiting is so familiar because it is the way that institutions make themselves known to us and because, in previous circumstances, waiting paid off. It is all this and more, but mainly, we wait because we have no choice.

In waiting we distribute our agency somewhere between here and now, here and there in a shared system of anticipation. David Bissell synopsizes the issue succinctly: 'Waiting for an event is a form of anticipation. A form of contractuality and a temporally displaced form of *trusting* relationality forged between subject and event-to-come' (Bissell 2007: 282). In this sense the traveller's experience seems to duplicate that of a person standing in a queue. Waiting for an event they cannot determine, they are becalmed by an anticipated directionality. Waiting in line or in the departure lounge, we are, in a sense, liberated from the present. We are already doing something by doing, apparently, not much. The relation between traveller and queuer is not just a structural likeness by dint of a shared distribution logic that mediates flows and blockages through linear sequencing of events, it is also a temporal relation that folds the passenger into probabilistic abstractions of the space of flows. The political dimension of anticipation is certainly not lost on those who wait. For instance, for Vladimir Sorokin, the soviet author of *The Queue*, standing in line was emblematic of a society governed 'by an ideology without a present tense; for society-in-waiting, kept in permanent thrall to the future' (Laird in Sorokin 1985: ii). For Walter Benjamin, 'he who waits' constituted a 'third type' of temporal relation between newly urbanizing man and his city. 'Finally, the third type: he who waits. He takes in the time and renders it up in altered form – that of expectation' (2004: 107).

This interest in waiting, this rendering up of time, is set in a context not only of informationalization, but also financialization of everyday life, where we mortgage time, energy and relationships for material security. A form of precarity underscores each as the probables of informational dynamics subdue the possibles of the virtual. In temporal spatial terms we might think a certain safety is offered by position and ordered sequence. Better to have a place in the line, than to be hurt in the crush, or worse yet, not be there at all.

Anticipate in an ancient sense was used in the context of considering something before *due* time. This looking forward to something creates an interesting interval in linear temporality as it causes the future event to be a reality beforehand. This future in present bumps the now into a type of future past that probabilizes what could happen and reacts pre-emptively. Standing in a now that is given over to the imagined future, we tingle with anticipation, or become lost in the risks that an unknown future could bring. Anticipatory architectures becalm the fear and feed hope.

The notion of anticipatory architecture can work in multiple ways. As a type of hypernetwork, airports have open architectures that extend and modify into each other. Their openness anticipates a future of constant development and adaptation. They are open to the unknown. However, managing traffic presses anticipation into the predictive register of logistics, whereby the techniques of parcelling, identifying and sequencing make components as interchangeable as possible, facilitating a smooth flow out of a field of potential relations. So, on one level the airport must constantly anticipate, not only traffic flows in order to allocate resources, such as gates, runways, flight paths in the most cost-effective manner, but also future trends and developments in aviation. Future upgrades are a constant of airport design. As a consequence, airports are endlessly renewing and rarely complete. On all levels, the airport is held in the thrall of what might be. Not just in terms of bigger jets or changing traffic patterns[3] but also in terms of potential threats – a bomb concealed in the heel of a shoe, a chemical cocktail transported in a baby bottle?

Anticipatory architecture focuses on relations with all manner of futures. In earlier times, the notion of architecture that 'looked forward' wasn't always geared towards the worst-case scenario, rather it inaugurated a potential utopian relationship between architecture and nature. For Manfredi Nicoletti, champion of 'continuity, openness and symbolisation',

> [a]rchitecture is acquiring meaning only through the interaction with the evolving ambience, through which a multiplicity of rapports, which is enlarging the limits of architecture to the whole environment, now understood as a continuum.
>
> (Nicoletti 1971: 279)

In a time of paranoia, anticipatory architectures structure relations of *mistrust* and pat you down twice, just in case. Within an informational milieu, we might articulate open architecture as plugging different components of an assemblage into different operating systems and channels of circulation, enabling different affectual relations and potentials. Anticipatory architectures structure emergent relations that modulate body, behaviour and social energies. How is this continuum of body to architecture managed in a context of fragmentation? Airports, for all their open architecture and seamless extension, are a series of corrals and gates – a series of pauses that are factored into the movement. Designed to store as well as forward, their architectures mesh capture and flow, and stillness and movement into a practical unity.

Figure 3.2 The airport's flow architecture materializes as a series of holdings as
well as movements.

A body between

As one sits watching the exercise video in the plane, circling one's ankles and
tensing one's toes in order to thwart potential deep vein thrombosis, it's hard
not to be acutely aware of the continuous relationship between bodies and
architecture. In this co-evolutionary relationship we have learnt to still
ourselves before images and wait in lines in other environments that converge
at the airport. In the departure lounge, we experience a convergence of the
panorama of cinema, as planes glide on the apron before us, and the intimacy
of the lounge room, as, surrounded by our most precious possessions, we watch
international news services on plasma screen TVs. One could say, as we are
stored, we also store ourselves.

For Sylvere Lotringer the question of Empire is, 'what is a body capable
of', or how to bolster one's own power (in Virno 2004: 7). For the 'kinetic
elite' endlessly circulating the space of flows – a business meeting in New
York – a teleconference at 10 a.m. Taiwan time – constant emailing with
London and so on – the body in question is multiple and fluid, but it is also,
on one level at least, also a body. I will speak here of a specific body, not
only as an antidote to the abstraction of the PAX, as discussed previously,
but also because my Antipodean body, which travels long distances regularly,
provides an extreme example of what the body endures in the air.

A body has its limits, and after thirty or so hours in various modes of stillness (sitting in the plane, the departure lounge, standing in various immigration, security and customs queues), nothing seems to 'organ'-ize the fluid, unfolding distributed self more than air travel. As a long-distance traveller I leave, an economy class cyborg loosening the pull of gravity, seeking connection and moving into new dimensions and arrive shattered – a laggy mess. The rapture of imminent release always involves capture of some kind. In order to travel, in order to deterritorialize, to move in a way that a human body normally couldn't, one must first become the most basic of bodies – a body with organs – a body that runs on metabolic time. This body needs sleep, food, air and a modicum of physical activity to keep the blood pumping. The limits of this body form its *locus operandi*; it can be sniffed, patted down, swiped, looked at and through. It can also be seated, routed and sequenced. With or without organs a body needs a body. It needs form. The 'unstable matters, the flows in all directions, the mad or transitory particles' that permeate the Body without Organs (Deleuze and Guattari 1998: 40) are simultaneously given form through stratification. Matter formed through what one might call productive limitation.

Yet these limits are also limits. For instance, a body has a complex chronobiology, from fast nervous rhythms to the slower beat of monthly menstruation. Our heads might be watching a rerun of Seinfeld on in-flight TV, our identity might be organized according to global logistics, but our body is experiencing the flight – its duration – in the first person. In aviation, this hybrid mix between first- and third-person time, between release and capture, deterritorialization and stratification, between the global timetables of work, play and movement and the metabolic rhythms of breath and heart is intense. The sometimes asynchronous and awkward relation between idealized, serial time, through which we attempt to organize our lives, and the embodied experience of duration crystallizes at the airport, where each of these times are multiple and embedded within each other in a series of part/whole relationships that are both integrated logistically as well as experienced pragmatically.

What happens when the actual body is neither here nor there (now nor then) but in constant transit, not only physically, but also behaviourally and cognitively? How do we manage the relations between interchangeable time, or what we might call serial time and non-reversible time or duration. In serial temporality, time is a slot that can be replaced by another. For instance, if I miss the 0900 to Melbourne, I can catch the 0930, and alternatively my slot on the 0900 can be exchanged with someone else. Moreover, while a jet-lagged body verifies the experience of duration in long-distance travel, and airports increasingly cater to it, in the form of sleeping rooms and massage services, duration is simultaneously abstracted away from the body and into the anticipatory realm. Time as duration is given over to delay, time as measure exists to estimate delays, 25 minutes, 4 hours. Linear time is given over to event time. One only moves in the line if the one in front moves. Your plane will land in a slot, and place becomes a position in a passage. This transitory dynamic extends fractally.

Perhaps we should return at this point to the act of being a passenger. The passenger's relation to transit is defined by their destination: in terms of the projected adventures they will experience in exotic locations offered through aeromobility; in terms of the clothes they pack and the money they take; or even in terms of what codes facilitate their traverse through the system. The quality of 'passage' is that *thereness* is part of *hereness*. Or, put more glibly, where one is going is always better than where one is when one is in the plane. The body declining in recline, the body unfolding into the atmosphere in the form of noxious carbon emissions, the body situated in multiple databases of state structures that can and will assemble identity according to whatever narrative delusion it is currently in the grip of. A body that can no longer keep up with itself, but is perpetually distracted by 'what is next' while waiting for 'something else'.

An airport is a 'macro' structure, not just in terms of its size but also in the sense that it automates particular procedures according to its social interpretation of mobility. The ordinary framing of global mobility reveals a quite clear social interpretation of mobility that, as others have noted, either ignores the integral role of stillness in motion systems, or positions it as anomalous and a problem to overcome.[4] In other words, stillness is a 'jam'. It is a boring queue. It is a time to do something else while you are waiting for something else. It takes time for the airport to process a passenger who is now thoroughly meshed in the multiple assemblages of aeromobility. So, while waiting at Singapore, you log onto your email. If being held within the logic of store forward architectures was once a kind of release from various work times, that is no longer so. There is no down time for the global kinetic and correspondingly little 'up time' for the globally restrained.

For the refugee, for the business traveller, for the temporary worker, waiting is a fact of life, a mode of life, waiting for something to change. For the traveller, there is an expectation (a confidence) that obedience to protocol will pay off. However, for the refugee or the temp worker, it probably won't. Time itself is stolen. Unable to plan, constantly on call, each day rolls into another waiting – a mode, which is neither here nor there; neither stop nor go, but restrained agitation towards movement. A moment to consider how many become one, how crowds becomes queues, and units becomes unities as the spatiotemporal coordinates of store-forward reinvokes the spectres of scarcity and the promises of abundance and a life suspended through anticipation – of those who wait. The juxtaposition, made by Augé (1995), Serres (1995)[5] and others of glittering towers and shanty towns, airports and refugee camps can be seen in the informational form in the topologies of waiting that occur at checkpoint, airport or on dial-up connection. If speed and movement are commodities, then delay is the control.

Next?

Mobile culture is a collective culture. Watching the multitudes circulate at an airport – some silent and singular – others huddling in groups – a few chatting or texting – one might think of the 'plurality of isolations' that for Sartre (2004) and so many others is an indicative by-product of the collectively synchronized life of a city. Today's city, plugged into the global space of flows, braids together many complex systems through its intrinsic ability to reconfigure and rearrange bodies and bits. In the supermodern informational city, you access email from your car and the subway has no cell dead spots; you can mesh. Always available, always on. Waiting for the next bit of data to trigger the next event.

Navigating through the various control structures of today's informational milieu, we are in a *Zu Befehl* state – we are 'like good soldiers . . . always in a state of conscious *expectation of commands*' (Canetti 1984: 312, my emphasis). For Canetti, 'a soldier is like a prisoner who has adapted himself to the walls enclosing him, one who does not mind being a prisoner and fights against his confinement so little that the prison walls actually affect his shape' (1984: 312). This affectual modulation shapes both system and user through a shared architecture of inputs and menus and commands that are so mutually implicative they are no longer generated outside (an *Ur*-principle of commands for Canetti), rather they take on the more intimate relations of prompts. The role of aeromobility is to flatten the affective tone of movement, limiting its potentials to forwardness, its focus to the next event. It doesn't take much thinking, sensing or doing to be 'in line'. Of course, anyplace where people congregate can form a 'community' of some sort; as stories told by friends and colleagues attest, the so-called 'terrible fact'[6] of the space of flows is that it is a place of sociality as well as security – a half-life of limiting but more certain potentials.

But it is more than this – in dealing with complex emergent systems (such as integrated global movement) there can be no absolute certainty about how anything will turn out. Anticipation – the suspenseful state – is a slippery and yet most effective mode of capture. We are suspended by the virtual temporality of anticipated events – we are after before and before after – our becoming bleeds into this abstraction – that of the next.

Modern logistical mobility is a mime of movement. It cuts movements into an infinite series of jerks. Each jerk suspends movement, and this suspense operates at all levels. For Brian Massumi, moments of suspense are where mind and potential merge (2002). What will happen next? For those of us who are global business travellers, we may not feel the jerks. And certainly, avionic architecture at all levels, from moving walkways to jet propulsion, works to inculcate a sense of smoothness and transparency in movement. It all seems so natural, so logical, as if it is happening all by itself. Mobility seems to be in perfect isomorphism with movement.

The trajectory of airports from the glamour of the Jet Age, in which airports were boutiques, to today's megastructures has been well documented

(Pascoe 2001; Pearman 2004; Fuller and Harley 2004). Airports now sport a familiar aesthetic wherever one is. This aesthetic articulates the anticipatory temporalities of networks, rather than the imagined glamour of jet set travel. The computer applications that model our lives, such as databases, simulations and visualizations have become cultural forms in the information society. Relations of filtering, screening, sorting, tagging, storing and forwarding are common across all network cultures and they permeate our consciousness as well as organize our lives. As Deleuze and Guattari note, 'it is now impossible not to identify with the network – one's intensive and extensive attributes reside there. They could not exist otherwise' (1998: 142).

'As life becomes more subject to administrative norms, people must learn to wait more' (Benjamin 2004: 119). This small notation, made by Benjamin in *The Arcades Project*, is easily overlooked among the intriguing aphorisms and the catchy quotes from Hugo and Baudelaire. Yet the relations between 'administrative norms' and 'learning to wait more', particularly in a context where 'time is money', could not be more intense or open to commercial and political exploitation. If part of the justification for moving to distributed control systems of information society and away from the temporal and spatial inefficiencies of centralized administration, what does it mean to wait now? How does one navigate the virally duplicating landscapes of transit, in which all interfaces are so familiar they are transparent?

> When there is no change in affective tone, things are transparent. You walk down the street without any sense of the fact that you are walking down the street. But at the moment the car honks the transparency is lost. You have to reinvent your behaviour.
>
> (Varela 2000:15)

At the airport, the aesthetics, rhetoric and operations of transparency work to ensure that you will not need to reinvent your behaviour. Breaks in transparency inaugurate moments where 'we realize that the now really exists' (Varela 2000: 16). Momentarily released from the future in present temporality and the daydreams we use to smooth the boredom of not being anywhere or anywhen, we may realize that, right now, I am being frisked or, right now, I am consciously contributing to global warming. And yet we wait. Or rather we move through waiting. A life of anticipation is the life of transit in which activity and passivity, stillness and motion, there and here mesh into a mobile ontology of a future tense.

Notes

1   Creswell's (2006) discussion on the Privium biometric system at Schiphol airport highlights the high uptake of the kinetic elite in this system of voluntary surveillance.
2   *The New York Times* (1/15/07) reports a rise in private jet services, such as onesky.com in which passengers pay premium prices for a flight 'so long as it does

not involve security lines, airport parking garages and competition for overhead baggage space' (Tedeschi 2007).
3   For instance some airports are now modifying gates and concourse for the introduction of the A380. The rise of China as a global commercial player has changed global traffic patterns and inaugurated an airport boom in China.
4   See for example, Bissell (2007) and Sofia (2000).
5   According to French anthropologist Marc Augé, today we live in a world where people are born in the clinic and die in the hospital, where transit points and temporary abodes are proliferating under luxurious or inhuman conditions (hotel chains and squats, holiday camps and refugee camps, shantytowns threatened with demolition or doomed to festering longevity; where a dense network of means of transport which are also inhabited spaces are developing; where the habitués of supermarkets, slot machines and credit cards communicate wordlessly, through gestures, with an abstract, unmediated commerce: a world thus surrendered to solitary individuality, to the fleeting, the temporary and ephemeral (1995: 78). See also Serres (1995).
6   Discussing Barcelona airport after it upgraded for the 1992 Olympics, Castells notes, 'In the middle of the cold beauty of this airport passengers have to face their terrible truth: they are alone, in the middle of the space of flows' (1998: 421).

Bibliography

Augé, M. (1995) *Non-places: An Introduction to an Anthropology of Supermodernity*, trans. J. Howe, London: Verso.
Benjamin, W. (2004) *The Arcades Project*, trans. H. Eiland and K. McLaughlin, Cambridge: Belknap Press of Harvard University Press.
Bissell, D. (2007) 'Animating Suspension: Waiting for Mobilities', *Mobilities*, 2(2): 277–98.
Canetti, E. (1984) *Crowds and Power*, New York: Farrar, Straus and Giroux.
Castells, M. (1998) *The Rise of the Network Society*, vol.1, Oxford: Blackwell.
Cresswell, T. (2006) *On the Move: Mobility in the Modern Western World*, New York: Routledge.
Deleuze, G. and Guattari, F. (1998) *A Thousand Plateaus: Capitalism and Schizo-phrenia*, trans. B. Massumi, Minneapolis: University of Minnesota Press.
Dodge, M. and Kitchin, R. (2004) 'Flying through Code/Space: The Real Virtuality of Air Travel', *Environment and Planning A*, 36: 195–211.
Fuller, G. and Harley, R. (2004) *Aviopolis: A Book about Airports*, London: Blackdog Publications.
Holmes, B. (2007) 'Future Map'. Available at brianholmes.wordpress.com/2007/09/09/future-map/ (accessed 1 November 2007).
Massumi, B. (2002) *Parables for the Virtual: Movement, Affect, Sensation*, Durham: Duke University Press.
Mau, B. and The Insitutute without Boundaries (2005) *Massive Change*, London: Phaidon.
Nicoletti, M. G. (1971) 'The End of Utopia', *Perspecta*, 13: 268–79.
Pascoe, D. (2001) *Airspaces*, London: Reaktion.
Pearman, H. (2004) *Airports: a century of architecture*, New York: Harry N. Abrams.
Sartre, J.- P. (2004) *Critique of Dialectical Reason*, vol. 1, trans. A. Sheridan Smith, London: Verso.

Schwartz, B. (1975) *Queueing and Waiting: Studies in the Social Organisation of Access and Delay*, Chicago: University of Chicago Press.

Serres, M. (1995) *Angels: A Modern Myth*, Paris: Flammarion.

Sofia, Z. (2000) 'Container Technologies', *Hypatia*, 15(2): 181–201.

Sorokin, V. (1985) *The Queue*, foreword by S. Laird, London: Readers International Inc.

Tedeschi B. (2007) 'For Sale: One-Way Trip on a Private Jet. No Waiting', *New York Times*, available at travel.nytimes.com/2007/01/15/technology/15ecom.html (accessed 10 November 2007).

Terranova, T. (2004) *Network Culture: Politics for the Information Age*, London: Pluto Press.

Varela, F. (2000) 'The Deep Now' in J.Brouwer and V2_Organization (eds.) *Machine Times*, Rotterdam: NAI Publishing.

Virno, P. (2004) *The grammar of the Multitude*, trans. I. Bertoletti, J. Cascaito and A. Casson, Cambridge: Semiotext(e).

Whitehead, A. (2004) *The Concept of Nature*, New York: Prometheus Books.

—— (1985) *Process and Reality* New York: The Free Press.

4 Connecting the world

Analyzing global city networks through airline flows

Ben Derudder, Nathalie Van Nuffel and Frank Witlox

Introduction

It has become commonplace to underline that recent developments in (the geographies of) transport and communication infrastructures have had a profound impact on the spatial organization of an increasingly globalized society (e.g. Black 2003; Rodrigue *et al.* 2006; Dicken 2007). One of the most commonly cited evolutions in this context is the alleged demise of the concept of 'territoriality' in favour of the concept of 'networks'. Leading sociologist Manuel Castells (1996; 2001) famously describes this as a transition from an international economy organized around 'spaces of places' to a global economy organized around 'spaces of flows'. Although there is a great deal of debate on the actual significance and implications of this shift, there can be little doubt that the spectacular growth of border-crossing mobility – for the largest part through air transport – is increasingly producing new spatial patterns of economic and social life. This has led some researchers to suggest that radical new ways of structuring our thinking about global spatial patterns are required, whereby a so-called 'global city network' (GCN) appears to be a likely candidate to provide the building blocks for such an alternative spatial framework (Taylor 2004a). Global cities are hereby essentially defined as key points in the organization of the global economy, and derive their functional importance from their mutual interactions rather than with their proper hinterlands.[1] In this chapter, we will focus on one particular aspect of the interrelation between transnational aeromobility and this networked meta-geography,[2] i.e. the relevance of data on air passenger flows for revealing the material spatiality of this GCN.

The chapter consists of three main parts. The first section presents a general consideration of the usefulness of airline data for analyzing GCNs. The relevance of airline data in this context seems deceitfully obvious: air transport is all about connections between cities, while airline data are comparatively easy to obtain. Our intention here, however, is to provide a somewhat deeper understanding of the relevance of airline data by situating this information source within the GCN literature at large. The second section shows how previous airline-based studies have quasi-systematically been hampered by inadequate and/or partial data. The third section, then, presents some possible alternatives to the problem of inadequate data. It is not our intention to provide yet another empirical analysis of global city-formation based on 'better' data, but rather to provide a conceptual overview of how airline-based analyses of GCNs may collectively be improved in future research. To this end, we discuss some alternative data sources and propose some data manipulations that, taken together, may advance our understanding of the relationship between aeromobility and GCNs. In a short conclusion, we briefly discuss the main implications of this chapter and outline some avenues for further research.

Air transport-based studies within global city network research

The contemporary GCN literature can be traced back to two interrelated papers by Friedmann and Wolff (1982) and Friedmann (1986). Both texts framed the rise of a global urban system in the context of a major geographical transformation of the capitalist world economy. This restructuring, most commonly referred to as the 'new international division of labour', was basically premised on the internationalization of production and the ensuing complexity in the organizational structures of multinational enterprises (MNEs). This increased economic-geographical complexity, Friedmann (1986) argued, requires a limited number of control points in order to function, and global cities were deemed to be such points. The publication of Saskia Sassen's (1991) *The Global City* in 1991 marked a shift of attention to global inter-city flows resulting from the critical servicing of worldwide production rather than to its formal command through corporate headquarters of MNEs. Sassen's approach focuses upon the attraction of advanced producer service firms (providing professional, financial and creative services for businesses) to major cities with their knowledge-rich environments and specialist markets. In the 1980s and 1990s many such service firms followed their global clients to become important MNEs in their own right. These advanced producer service firms thereupon created worldwide office networks covering major cities in most or all world regions, and it is exactly the myriad of interconnections between service complexes that, according to Sassen (1991; 2001), gives way to GCN formation.

Empirical GCN research has long been underdeveloped because of the lack of appropriate data, a problem that Short *et al.* (1996) referred to as 'the dirty little secret of world cities research'. This empirical poverty can, for instance, clearly be read from Castells' (1996: 469) book, which is part of a trilogy that is above all an attempt to reformulate social studies for a global age in which 'networks constitute the new social morphology of our societies'. However, when it comes down to providing a basic cartography of this global network society, Castells' argument falls short of the conceptual shift he advances: the only actual evidence he comes up with in the chapter on the 'space of flows' consists of some limited inter-city information gathered from Federal Express.

One can therefore only conclude, as Taylor (2004a: 35) has recently done, that 'the evidence [Castells] marshalls is mightily unimpressive'. This gap between theoretical sophistication and evidential poverty is, however, not a lacuna specific to Castells' book: it has been a *structural* feature of research on the GCN, because data for assessing such urban networks are in general insufficient or even totally absent.

The basic reason for this problem of evidence is that standard data sources are ill-suited for GCN analyses (Taylor 1997; 2004a). To get an evidential handle on big issues, researchers normally rely on the statistics that are available, that is to say, already collected. But such collection is carried out usually by a state agency for the particular needs of government policy rather than for social science research. The result is that such data that are available have an attributional bias (measurements *of* administrative areas rather than *between* administrative areas) and are limited to national territories. Where official statistics extend beyond a state's boundaries, they will still use countries as the basic units (e.g. trade data). Thus there is no official agency collecting data on, say, the myriad flows between London and New York. The major result has been that 'few of the available data reveal anything about the flows and interdependencies' that are at the heart of this body of literature (Knox 1998: 26), which leads Alderson and Beckfield (2004: 814) to note that in the past relatively few of the empirical GCN studies 'utilized the sorts of relational data necessary for firmly establishing such rankings empirically'.

These data problems have put researchers to work in recent years, and we have therefore witnessed a proliferation of empirical studies that explicitly seek to rectify this situation. Researchers have hereby relied on a wide variety of data, albeit that some information sources have come to dominate the empirical research as a whole (Derudder 2006), i.e. (i) information on corporate organization (e.g. data on ownership links between firms across space) and (ii) information on infrastructure networks (e.g. data on the volume of air passenger flows across space). The success of both approaches can, of course, be traced back to their commonsensical appeal: the corporate organization approach acknowledges that well-connected cities derive their status in large part from the presence of key offices of important firms, while the infrastructure approach recognizes that well-connected cities are typified by the presence of vast enabling infrastructures. Put simply: the most important cities harbour the most important airports, while the extensive fibre backbone networks that support the Internet have equally been deployed within and between major cities, hence creating a vast planetary infrastructure network on which the global economy has come to depend almost as much as on physical transport networks (Rutherford *et al.* 2004).

Table 4.1 summarizes the approaches developed in the empirical GCN literature through an overview of some key studies in this research domain. The table acknowledges that the basic bifurcation between corporate organization and infrastructure needs to be deepened on the basis of the exact types of firms and infrastructure, and equally shows that all this is in practice

Table 4.1 A taxonomy of empirical approaches

Indicators	Corporate organization		Infrastructure		Other and/or a combination of indicators
	Global service firms	Multinational enterprises	Telecommunications	Air transportation	
Examples	Beaverstock et al. (2000a)	Alderson and Beckfield (2004, 2007)	Townsend (2001a;b)	Keeling (1995)	Beaverstock et al. (2000b)
	Taylor et al. (2002)	Rozenblat and Pumain (2007)	Malecki (2002)	Smith and Timberlake (2001, 2002)	Taylor (2004b)
	Derudder et al. (2003)		Dupuy (2004)	Cattan (1995, 2004)	Rimmer (1998)
	Derudder and Taylor (2005)		Rutherford et al. (2004)	Matsumoto (2004, 2007)	
				Zook and Brunn (2005)	

Source: Derudder 2006.

somewhat more complicated because of the presence of a limited number of studies that (i) make use of other types of datum (e.g. Taylor's (2004b) analysis of non-governmental organizations) and/or (ii) combine indicators from both approaches (e.g. Beaverstock *et al.* 2000b). In the next section, we focus on empirical GCN studies that utilize data on international aero-mobility to map GCNs.

Data issues in airline-based GCN studies

Preamble

The starting point of airline-based GCN studies is the rather commonsensical observation that interactions between global cities are in large part facilitated and defined by transnational aeromobility. Following Keeling's (1995) initial contribution, there have been a large number of empirical researches that draw upon airline data to devise a mapping of the GCN (e.g. Cattan 1995; 2004; Short *et al.* 1996; Kunzmann 1998; Rimmer 1998; Shin and Timberlake 2000; Smith and Timberlake 2001; 2002; Matsumoto 2004; 2007; Zook and Brunn 2005). In principle, the most important advantage of this approach over researches carried out in the corporate organization approach is that airline statistics feature *tangible* inter-city relations. However, in hindsight, and in spite of the remarkable success of this type of research, it can be noted that most authors have simply asserted the relevance of publicly available airline data, although these are – in our view – downplayed by a number of structural problems (for earlier, but partial, assessments, see Taylor (1999) and Beaverstock *et al.* (2000a,b)). In this section, we will provide a systematic overview of these data problems, which will in turn be used in the next section to show how this baleful situation may be rectified.

The first problem: the lack of origin/destination data

In spite of profound differences between the most commonly employed air transport statistics – i.e. those provided by the Official Airline Guide (OAG), the International Air Transport Association (IATA) and the International Civil Aviation Organization (ICAO) – these data sources are collectively hampered by at least three problems. A first major problem is the *lack of origin/destination information*. Standard airline statistics feature the individual legs of trips rather than the trip as a whole. Thus, in the case of a stopover, a significant number of 'real' inter-city links are replaced by two or more links that reflect corporate strategy rather than GCN relations. Furthermore, the lack of origin/ destination information makes geographically detailed assessments of the GCN difficult, as direct connections become less likely as one goes down the urban hierarchy. And finally, the emergence of hub-and-spoke strategies in the airline sector have led to the rise of a set of urban areas that serve as inter-connection points between different regions (Bowen 2002; Derudder *et al.*

2007). As a consequence, there is a continuous shift towards a more polycentric organization of airline passenger networks and away from the traditional orientation on major cities. This implies that a number of secondary cities are rapidly gaining prominence in this new polycentric structure because of their role as transfer points rather than as origins and/or destinations in their own right, and the way in which airline statistics are commonly recorded tends to give disproportionate weight to such cities in GCN analyses.

An example of this first data problem can be observed in Keeling (1995), in which the mapping of the GCN is derived from an analysis of the dominant linkages in the global airline network. This map was created from a matrix of scheduled air services between 266 cities, based on OAG data. However, this implies that only non-stop and scheduled direct flights between two cities were taken into account. As a consequence, the measures used by Keeling are not necessarily a reflection of actual inter-city relations. That is, such an analysis is likely to overstate the relational importance of cities that function as airline hubs, such as Amsterdam (KLM) and Frankfurt (Lufthansa), at the expense of cities such as Brussels and Berlin. Furthermore, direct links between, say, Brussels and Rio de Janeiro cannot be measured, as passengers are likely to go through São Paulo to make this trip.

The second problem: state-centrism in data

The second obstacle to translating air transport statistics into GCN analyses arises because some of these data sources have incorporated *a subtle form of state-centrism*. Despite their global aspirations, most analyses are based on databases that contain information on international flows. This bias towards inter-state rather than trans-state flows tends to undervalue relations between cities that are situated in large and/or significant nation-states. Rimmer (1998: 460), for instance, has based his GCN analysis on data on 'international passengers'. This results in a downgrading of US world cities in particular, because important connections such as Los Angeles–New York and Chicago–New York are not incorporated in this approach. As a consequence, Chicago only appears on one of Rimmer's maps as a 'fourth-level' link to Toronto, while Dublin appears on all maps because of its 'first-level' link with London. Of course, nobody would argue that Dublin is more important than Chicago as a global city; it only appears to be this way when one relies on international rather than transnational data. Another example can be found in Smith and Timberlake (2002: 123), who report a lack of information on the volume of air passenger traffic between Hong Kong and London. This admittedly important inter-city link did not feature in pre-1997 databases of the ICAO because flights between London and Hong Kong were considered 'national'. While the classification of the London–Hong Kong route and the downgrading of US cities are extreme examples, they clearly reveal how data on international passenger flows may hamper a GCN analysis.[3]

The third problem: general flow patterns

The third obstacle to the straightforward use of standard air transport statistics arises from the fact that such data sources feature *general flow patterns*. Since airline statistics are unable to differentiate between specific flows within air passenger transport (i.e. the purpose of a passenger's travels), it is doubtful that aeromobility in the context of the GCN can straightforwardly be deduced from such general data. A key example is the inclusion of major tourist destinations in previous analyses. For instance, in his mapping of the European urban system based on air passenger flows, Kunzmann (1998: 49) lists fourteen airports that are secondary to the big three (London, Paris and Frankfurt), including Munich, Milan, Madrid and Palma de Mallorca. However, the high ranking of the latter merely reflects its role as one of the most popular holiday destinations in Europe; nobody would argue that it is a major global city (however conceived). While it is likely that most researchers would agree that destinations such as Palma de Mallorca should be omitted from the analysis, such data manipulation becomes increasingly difficult when non-global city processes intersect with global city-formation. The rising importance of Miami in airline networks, for instance, can in part be traced back to its rise as control centre vis-à-vis the Carribean (Grosfoguel 1995; Nijman 1997; Brown *et al.* 2002). However, it is obvious that in the main it has been Miami's function as a retirement centre and major holiday destination that has fuelled its increasing connectivity in worldwide air transport connections. Since none of the commonly employed airline statistics are able to distinguish between tourist and business flows, there have been no clear procedures for estimating the amount of GCN-related traffic in overall air travel.

Towards some solutions

In the previous section, we have argued that the key advantage of air transport-based studies (i.e. their focus on tangible inter-city flows) is downplayed because standard air transport statistics are collectively plagued by three deficiencies, i.e. (i) the lack of origin-destination data, (ii) the state-centric nature of the most important airline statistics and (iii) the intersection with non-GCN processes. In this section, we show how future research may overcome these problems by making use of alternative datasets, combined with some specific manipulations of these data. We will show how the first two problems can be tackled by using a so-called marketing information data transfer (MIDT) database, while the third problem can be addressed – at least for Europe – by drawing on a dataset of the Association of European Airlines (AEA) that distinguishes between different travel classes.

MIDT data

Our first alternative data source comprises a unique data set (for social science research) that provides information on individual passenger flows in

2001. This MIDT database is described in detail in Derudder and Witlox (2005) and Derudder *et al.* (2007), and reference should be made to these publications for further details. Here we produce a summary so that our line of reasoning can be followed. The MIDT database contains information on bookings made through so-called global distribution systems (GDSs) such as Galileo, Sabre, Worldspan, Topas, Infini, Abaccus and Amadeus (Shepherd Business Intelligence, 2005). GDSs are electronic platforms used by travel agencies to manage airline bookings (i.e. the selling of seats on flights offered by different airlines), hotel reservations and car rentals. Using a GDS-based database therefore implies that bookings made directly with an airline are often excluded from the system and therefore the data. However, in 1999, just two years prior to our data, 80 per cent of all reservations continued to be made through GDS (Miller 1999). Thus, although our information source may give a slightly biased picture of airline connections, there is no reason to assume that the overall pattern of reservations made by direct bookings differs fundamentally from that for reservations made through a GDS.

Using this MIDT database instead of standard data sources has two conceptual advantages in the context of GCN research. First, as the MIDT database contains real origin/destination information, the overrating of the connectivity of airline hubs and first-tier global cities is minimized, which allows flows between cities in the lower rungs of the GCN to be assessed in more detail. Second, the MIDT-based database does not distinguish between national and international flows, and can therefore be used to construct a truly transnational inter-city matrix. The New York–Chicago link is appropriately treated in the same way as the New York–Toronto link, which further reduces the underestimation of second-tier cities in large and/or significant nation-states.

Through our cooperation with an airline, we were able to obtain an MIDT database that covers the period January–August 2001 and contains information on more than 500 million passenger movements. This database was used to construct an inter-city matrix detailing the total volume of passenger flows between cities. To achieve this, we first relabelled the airport codes as city codes. This was necessary to compute meaningful inter-city measures, because a number of cities have more than one major airport. The particular airport used by a passenger is not important in this context because, for the measurement of the London–New York relation, it is irrelevant whether one flies from Heathrow to JFK or from Gatwick to Newark. Having summed the directional information into a single measurement detailing the total volume of passengers, we created an inter-city matrix that focuses on the most important cities in the global economy. Our selection of cities is based on the global city list compiled by the Globalization and World Cities (GaWC) group and network (www.lboro.ac.uk/gawc/datasets/da11_2.html). Nine of the initial 315 cities were excluded because they had no airport (e.g. Bonn and Kawasaki) or because the airport was not serviced in the period under consideration because of political instability (e.g. Kabul). This reconfiguration produced a 306 × 306 matrix that quantifies the passenger flows between important cities in the global economy. Table 4.2 and Figure 4.1 present an overview of the most important cities and their connections. Table 4.2 features the twenty most important inter-city relations in the dataset. Figure 4.1 depicts the connections between the

Table 4.2 Most important inter-city links in the MIDT database

Rank	Connection		Number of passengers
1	Hong Kong	Taipei	2,138,484
2	Los Angeles	New York	1,607,593
3	London	New York	1,609,337
4	Melbourne	Sydney	1,563,106
5	Milan	Rome	1,518,767
6	Cape Town	Johannesburg	1,406,897
7	Los Angeles	San Francisco	1,375,660
8	Amsterdam	London	1,242,550
9	Chicago	New York	1,182,326
10	Bangkok	Hong Kong	1,141,062
11	London	Paris	1,060,999
12	Dublin	London	1,050,940
13	Marseilles	Paris	1,044,128
14	Bangkok	Singapore	1,024,818
15	Rio De Janeiro	São Paulo	992,775
16	Boston	New York	988,976
17	Miami	New York	955,838
18	Atlanta	New York	935,265
19	Las Vegas	Los Angeles	924,732
20	New York	San Francisco	909,514

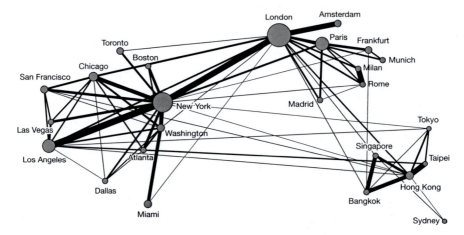

Figure 4.1 Most important cities and links in the world city network (based on MIDT, 2001).

thirty most important cities in terms of total passenger flows. The size of the nodes varies with the total number of incoming or outgoing passengers; the size of the edges varies with the number of passengers flying between two cities. For reasons of clarity, only the most important links are shown.

To illustrate the basic differences between our MIDT dataset and more conventional data sources, Table 4.3 presents a comparison of the twenty most connected cities in the airline network as identified (i) in the ICAO-based analysis of Smith and Timberlake (2001; 2002) and (ii) in our MIDT database.[4] Two differences between these rankings stand out. First, cities with a lot of hub traffic are less important in the MIDT ranking (e.g. Paris, Amsterdam and Frankfurt). This is, however, not the case for major US hubs (e.g. Atlanta and Dallas), because this first tendency is offset by a second difference between

Table 4.3 Comparison of most important cities in Smith and Timberlake and the MIDT database

Rank	Smith and Timberlake (2001; 2002)	Rank	MIDT database
1	London	1	New York ↑
2	Frankfurt	2	London ↓
3	Paris	3	Los Angeles ↑
4	New York	4	Paris ↓
5	Amsterdam	5	Chicago ↑
6	Miami	6	San Francisco ↑
7	Zurich	7	Hong Kong ↑
8	Los Angeles	8	Miami ↓
9	Hong Kong	9	Frankfurt ↓
10	Singapore	10	Atlanta ↑
11	Tokyo	11	Madrid ↑
12	Seoul	12	Toronto ↑
13	Bangkok	13	Washington ↑
14	Madrid	14	Rome ↑
15	Vienna	15	Bangkok ↓
16	San Francisco	16	Milan ↑
17	Chicago	17	Amsterdam ↓
18	Dubai	18	Dallas ↑
19	Osaka	19	Boston ↑
20	Brussels	20	Singapore ↓

Source: Smith and Timberlake 2001; 2002; Derudder and Witlox 2005.

both rankings. That is, because of the transnational basis of the MIDT data (e.g. the New York–Chicago link is appropriately treated in the same way as the New York–Toronto link), US cities tend to have higher connectivities at the expense of Southeast Asian hub cities (e.g. Dallas and Atlanta versus Osaka and Singapore).

AEA data

The main problem with the MIDT dataset is that it remains largely impossible to discern GCN-related flows from other flows. The importance of the New York–Miami route and particularly the New York–Las Vegas route in an overview of the most important North American inter-city links, for instance, suggests the importance of non-GCN links in the data (Table 4.4). Flows related to obvious holiday destinations such as Palma de Mallorca and Cancún can easily be deleted from the database, but this manipulation only works for airports that are obviously *not* related to global cities.[5]

To counter this problem, one may draw upon data such as those provided by the AEA to unveil the basic spatiality of the GCN (albeit in this particular case only within Europe). The AEA is a non-profit-making organization that brings together thirty-one major European airlines. The organization represents its member airlines in dialogue with all the relevant European and international organizations in the aviation value chain, thus ensuring the sustainable growth of the European airline industry in a global context. The AEA gathers travel data among its member airlines and brings this information together in a database that allows the geography of airline networks serviced by major European airlines to be assessed on a monthly basis. For each connection, the database features information on carrier, origin and destination (airport, city, country and region), and the total volume of passengers, freight and mail. The AEA database seems very suitable for research on GCNs, because the informa-tion on passenger volumes distinguishes between economy class and business

Table 4.4 Most important North American inter-city links in the MIDT database

Rank	Connection		Number of passengers
1	Los Angeles	New York	1,697,593
2	Los Angeles	San Francisco	1,375,660
3	Chicago	New York	1,182,326
4	Boston	New York	988,976
5	Miami	New York	955,838
6	Atlanta	New York	935,265
7	Las Vegas	Los Angeles	924,732
8	New York	San Francisco	909,514
9	Las Vegas	New York	856,221
10	New York	Washington	806,875

class bookings. As a consequence, the data allow for an exclusive focus on the geography of business class travel (which will very likely be a better measure of GCN connectivity) and an assessment of the way in which this geography differs from that produced by 'ordinary' air passenger flows.[6]

Through the cooperation of one the member airlines of the AEA, we obtained this dataset for the period January 2002 to December 2005. The database includes flights within Europe as well as flights between Europe and other regions. For our purposes, only flights where both origin and destination are European cities were retained. Furthermore, since our interest is primarily in flows between cities, we converted the airport-to-airport-by-carrier database into a squared city-to-city database by aggregating, for each booking class, the number of passengers for all the airports of a given city, and for all the carriers of a given city-pair. The end result, then, is a non-directional connectivity matrix for each type of booking class.[7]

The most important problem with this and other similar datasets lies in the disparities in business class bookings because of different strategies pursued by 'national carriers'. This bias relates to the fact that most airports are (still) dominated by one or two carriers, which may or may not have a specific approach towards business class travel. Some carriers, for instance, have recently chosen to remove business class from some of their short/medium-haul routes (e.g. KLM and Brussels Airlines). On these routes, the seats are the same for all passengers; only the flexibility of the ticket and the food and beverage service differ. On the other side of the spectrum, some carriers have a relatively important business class component because they serve specific markets (e.g. carriers flying to London City Airport) and/or because of historic reasons. The latter is the case for travel to/from Scandinavian cities. SAS, the dominant regional carrier in Scandinavia, was the first European airline to introduce business class, and this is still reflected in a high proportion of business class seats on their flights. The net effect of this bias is that business class bookings for, say, Copenhagen and Stockholm will be somewhat overvalued when compared with, say, Brussels and Amsterdam, while the connectivity of cities that are located near other important business centres will be somewhat undervalued because business travellers can choose between different means of travelling.

One further aspect of this dataset requires explicit interrogation, i.e. the crucial issue of whether business class bookings actually capture the spatiality of business travel (and therefore GCN flows). After all, business travellers do not necessarily travel in business class, while (ostensibly rich) tourists may well travel in business class because of enhanced comfort. Thus the importance of, say, New York's GCN flows may well be overestimated because of rich leisure travellers, while business flows to some short-haul destinations may well be underestimated because business travellers may opt not to travel in business class because of the short travel time. The crucial question, therefore, is whether measures of *business class travel* provide us with satisfactory proxies for assessments of *business-related travel*? At one level distortions are clearly present, but the crucial question here is whether the ensuing biases are so strong that they totally undermine an analysis of business travel on the basis of business class bookings.

To address this issue, we discuss two basic features of the AEA dataset for the year 2005, which jointly suggest that data on business class travel do indeed allow an actual assessment of business travel. First, there seems to be a straightforward difference in seasonal intensity for both types of booking class. Figure 4.2 gives an overview of the monthly fluctuations in air travel in 2005 for the entire AEA database for both booking classes. The monthly variations in connectivity are gauged through z-scores, so that inter-booking class comparisons are possible in spite of different passenger volumes. The seasonality of air travel is obviously different for economy and business class bookings. The economy class curve increases from January to July/August, and then decreases again towards the end of the year. The business class curve, in contrast, reaches its lowest levels in major holiday periods (July/August and December/January). The major point here is that the contrasting curves in Figure 4.2 suggest that, in general, air travel in business class captures business travel.

Second, the relative proportion of business class travel is higher for clear-cut 'business cities' such as Geneva and Düsseldorf. Table 4.5 contains two rankings of European cities according to their connectivity in the European airline network. The first ranking focuses on the *absolute* importance of business class travel, the second ranking focuses on the *relative* proportion of business class travel within a city's overall passenger volume. When taking on board that (i) the proportion of business class travellers to/from Scandinavian cities is higher because of the historical legacy of SAS's corporate strategies, and that (ii) the proportion of business class travel to/from cities such as London, Paris, Frankfurt and Amsterdam is somewhat relegated

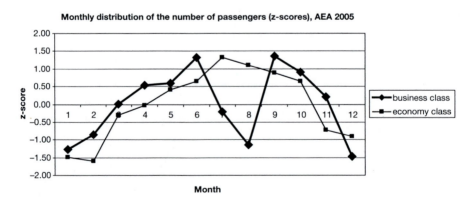

Figure 4.2 Monthly distribution of the number of passengers through z-scores (based on AEA, 2005).

Table 4.5 Banking of European cities according to their business class connectivity in the AEA database (2005)

		Total business			Proportion business
1	London	3,281,117	1	Geneva	16.26
2	Frankfurt	2,026,604	2	Oslo	16.00
3	Paris	2,019,845	3	Stockholm	14.78
4	Amsterdam	1,737,635	4	Düsseldorf	14.62
5	Copenhagen	1,096,543	5	Frankfurt	13.61
6	Munich	1,080,402	6	London	12.95
7	Stockholm	863,045	7	Zurich	12.77
8	Milan	853,438	8	Copenhagen	11.80
9	Vienna	764,851	9	Munich	11.75
10	Brussels	763,111	10	Vienna	10.75
11	Madrid	759,496	11	Brussels	10.72
12	Oslo	694,255	12	Amsterdam	10.10
13	Geneva	568,867	13	Paris	9.67
14	Rome	514,360	14	Milan	9.32
15	Düsseldorf	461,820	15	Berlin	8.39
16	Prague	419,508	16	Madrid	8.24
17	Zurich	413,365	17	Prague	8.03
18	Barcelona	397,845	18	Helsinki	7.62
19	Istanbul	394,400	19	Athens	7.21
20	Athens	338,843	20	Manchester	7.18
21	Helsinki	332,139	21	Istanbul	7.09
22	Lisbon	310,424	22	Budapest	7.09
23	Budapest	256,799	23	Rome	6.77
24	Hamburg	219,458	24	Lisbon	6.59
25	Manchester	202,300	25	Barcelona	6.16

because of their function as gateways for rerouting international air travel (Derudder *et al.* 2007), it becomes clear that business centres do in effect have a higher proportion of business class travel. Once again, this seems to validate our claim that business class travel does indeed measure business travel.

To conclude this section, we extend the basic rankings in Table 4.5 by focusing on one further aspect of the geography of business travel. First, rather than restricting the discussion to the absolute and relative dimensions of business travel on a city-by-city basis, we can assess the actual spatiality of business flows between cities. To this end, Figure 4.3 depicts the most important business travel links in 2005 between the twenty-five most important European cities in terms of total volume of business passengers. Once again, the size of the nodes varies with the total number of incoming or outgoing passengers, while the size of the edges varies with the number of business passengers flying between two cities. For reasons of clarity, only the most important links are shown (> 50,000 passengers). In addition to a cohesive business network centred on Stockholm, Oslo and Copenhagen, it is clear that business travel

Table 4.6 Least squares regression on business and economy class flows in the AEA database (2005): standardized residuals larger than one standard deviation

London		Paris		Frankfurt		Amsterdam	
City		*City*		*City*		*City*	
positive		*positive*		*positive*		*positive*	
Geneva	3.09	London	3.59	Brussels	2.92	London	5.16
Frankfurt	3.05	Amsterdam	3.30	Zurich	2.86	Paris	3.39
Düsseldorf	2.67	Geneva	2.52	Geneva	2.50	Frankfurt	1.45
Brussels	1.76	Frankfurt	1.71	Basle	1.30		
Zurich	1.01	Düsseldorf	1.18	Milan	1.24		
				Amsterdam	1.00		
negative		*negative*		*negative*		*negative*	
Barcelona	−2.02	Barcelona	−1.59	Barcelona	−1.55	Barcelona	−1.29
Rome	−1.88	Madrid	−1.46	Istanbul	−1.45		
Dublin	−1.74	Rome	−1.22	Madrid	−1.44		
Lisbon	−1.35			Rome	−1.41		
				Lisbon	−1.19		

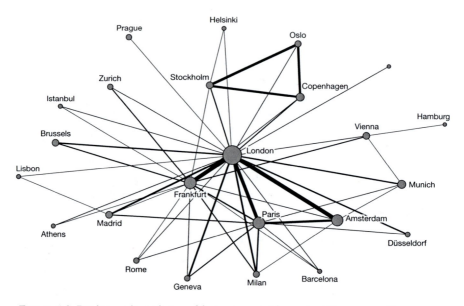

Figure 4.3 Business class air travel between most important European cities (AEA, 2005).

to/from Frankfurt, London, Paris and Amsterdam is dominant. These cities are highly interrelated, while most business travel from/to other major cities is also primarily orientated towards these cities (e.g., each city has well-connected business class flows to London). Table 4.6, in turn, looks at the spatiality of the relative importance of business travel to/from the most important business centres in 2005 (London, Paris, Frankfurt and Amsterdam). The table summarizes the results of a least squares regression on the logarithms of the volume of economy and business class passengers to/from each of these cities,[8] and lists all cities with a standardized residual with an absolute value larger than 1: large negative residuals indicate that a city has fewer business class travellers than expected on the basis of the number of economy class passengers; positive residuals point to relatively strong business class connections to London, Paris, Frankfurt and Amsterdam. Overall, the table reveals that cities with large positive residuals are primarily business centres (e.g. Frankfurt and Zurich), while cities with negative residuals are cities that also have an important tourism component to them (e.g. Rome and Barcelona).[9]

Conclusion

The gist of our argument has been that, although statistics of international aeromobility are potentially a prime data source in empirical studies of GCNs, previous such studies have not been able to live up to their potential. This is because the conventional airline statistics provided by IATA, OAG, ICAO, . . . result in less-than-perfect inter-city matrices for further analysis. We identified three systematic data problems in this context, and used this as a starting point for exploring alternative datasets. We discussed two such alternatives (MIDT and AEA), which jointly open up possibilities for future research along these lines. However, this chapter merely engaged in a straightforward overview of some of the major patterns in the data. In the short run, this leaves us with two major avenues for further research.

First, there is the question of how both data sources may inform one another. One obvious problem is that both datasets offer partial solutions (the MIDT data featuring general flow data, and the AEA data featuring the individual legs of a trip): it is only through their collective application that we can take full advantage of their potential. Leaving aside the obvious problem that the AEA dataset only contains information on European carriers (similar 'regional' datasets will be necessary for a global analysis), this will take the form of a normalization of MIDT data based on the proportion of business travel for particular cities/connections in the AEA data. Second, there is the actual analysis of the data. One possibility is to analyze the data along the lines suggested by Shin and Timberlake (2000). This involves a two-step analysis, i.e. (i) an overarching analysis of the overall connectivity of key cities and (ii) a detailed geographical dissection of this connectivity through the application of standard network analytical tools, such as clique and block analyses.[10]

Notes

1 Some researchers differentiate between the terms 'global city' and 'world city' (e.g. Sassen 2001), and in some cases such distinction is no less than crucial (see Derudder 2006). However, in the context of the present paper, these conceptual details are of lesser importance, and we will therefore consistently use the generic term 'global city' to address the literature at large.

2 In this paper, the term 'transnational' is used as a generic notion referring to the terms 'international' and 'intra-national' at the same time. Thus the concept of 'transnational aeromobility' refers to airline flows irrespective of the territorial framework in which this aeromobility occurs (i.e. between or within countries).

3 Smith and Timberlake (2002) were able to overcome the London–Hong Kong problem by estimating the importance of this link. The relegation of US cities was dealt with by the use of an additional data source that contained information on major routes in the US (namely, data provided by the Air Transport Association in Washington DC). While circumventing the most obvious gaps in the initial database has produced one of the most refined databases used to date, in general, this problem continues to affect the major Canadian, Chinese and Brazilian cities (among others).

4 It is important to stress that this comparison is only made for illustrative purposes. The list of 'most important cities' in Smith and Timberlake (2001; 2002) is constructed on the basis of a fully fledged centrality analysis, while the MIDT list merely reflects a ranking based on the total number of passengers boarding on/off. Furthermore, the data in Smith and Timberlake (2001; 2002) refer to passenger flows in the year 1997, while the MIDT data feature statistics for the year 2001. As a consequence, the reader should bear in mind that we merely compare both lists to point to some data-induced differences.

5 Another, more generic problem is that airline data cannot avoid undervaluing a second-tier city that is close to a major world city. For example, a passenger travelling from Rotterdam to New York is likely to depart from Amsterdam because of (i) the short distance between Rotterdam and Amsterdam (less than 50 miles) and (ii) the importance of Amsterdam's Schiphol airport.

6 The AEA dataset equally includes information on first class passengers. However, the absolute volume of these flows is very small for intra-European flights, and furthermore restricted to a select number of carriers. We have therefore chosen to add these flows to the business class category.

7 Once again, a further data manipulation involved the removal of obvious holiday centres such as Palma de Mallorca from the data through the application of GaWC's global city list.

8 The logarithms were used to tackle the heteroscedasticity in the data.

9 Obviously, the latter cities are also business centres in that they are also highly connected in business class flows (see Figure 4.3), but the point here is that their business class component is coupled with major tourism flows, which reinforces our suggestion that data on business class travel can actually be used to assess the geography of business travel.

10 Clearly, the MIDT and AEA databases also open up possibilities for research beyond the narrow confines of the formal empirical measurement and analysis of GCNs. Indeed, it is possible to imagine wider benefits from using the databases presented in this chapter. For instance, one might envisage a more refined analysis of the geography of global tourism by combining both datasets: the results of the above-mentioned normalization procedure of MIDT data based on the proportion of business travel for particular cities/connections in the AEA data can be analyzed from the observed perspective. This would result in an overview of which cities/connections are dominated by non-business flows.

Bibliography

Alderson, A. S. and Beckfield, J. (2004) 'Power and Position in the World City System', *American Journal of Sociology*, 109: 811–51.

—— and —— (2007) 'Globalization and the World City System. Preliminary results from a longitudinal data set', in P. J. Taylor, B. Derudder, P. Saey and F. Witlox (eds.) *Cities in Globalization: Practices, Policies, Theories*, London: Routledge: 21–36.

Beaverstock, J. V., Smith, R. G. and Taylor, P. J. (2000a) 'World City Network: a New Metageography?', *Annals of the Association of American Geographers*, 90: 123–34.

Beaverstock, J. V., Smith, R. G., Taylor, P. J., Walker, D. R. F. and Lorimer, H. (2000b) 'Globalization and World Cities: Some Measurement Methodologies', *Applied Geography*, 20: 43–63.

Black, W. R. (2003) *Transportation: A Geographical Analysis*, New York: Guilford Press.

Bowen, J. (2002) 'Network Change, Deregulation, and Access in the Global Airline Industry', *Economic Geography*, 78: 425–40.

Brown, E., Catalano, G. and Taylor, P. J. (2002) 'Beyond World Cities. Central America in a Space of Flows', *Area*, 34: 139–48.

Castells, M. (1996) *The Information Age: Economy, Society and Culture, Volume I – The Rise of the Network Society*, Oxford: Blackwell.

—— (2001) *The Rise of the Network Society,* 2nd edn, Oxford: Blackwell.

Cattan, N. (1995) 'Attractivity and Internationalisation of Major European Cities: the Example of Air Traffic', *Urban Studies*, 32: 303–12.

—— (2004) 'Le monde au Prisme des Réseaux Aériens', *Flux*, 58: 32–43.

Derudder, B. (2006) 'On Conceptual Confusion in Empirical Analyses of a Transnational Urban Network', *Urban Studies*, 43: 2027–46.

—— Devriendt, L. and Witlox, F. (2007) 'Flying where you don't want to go: an Empirical Analysis of Hubs in the Global Airline Network', *Tijdschrift voor Economische en Sociale Geografie*, 98: 307–24.

—— and Taylor, P. J. (2005) 'The Cliquishness of World Cities', *Global Networks*, 5: 71–91.

—— and —— Witlox, F. and Catalano, G. (2003) 'Hierarchical Tendencies and Regional Patterns in the World City Network: A Global Urban Analysis of 234 Cities', *Regional Studies*, 37: 875–86.

Derudder, B. and Witlox, F. (2005) 'An Appraisal of the Use of Airline Data in Assessing the World City Network: a Research Note on Data', *Urban Studies*, 42: 2371–88.

Dicken, P. (2007) *Global Shift: Mapping and Changing the Contours of the World Economy*, 5th edn, New York: Guilford.

Dupuy, G. (2004) 'Internet: une Approche Géographique à l'Echelle Mondiale', *Flux*, 58: 5–19.

Friedmann, J. (1986) 'The World City Hypothesis', *Development and Change*, 17: 69–83.

—— and Wolff, G. (1982) 'World city formation: an agenda for research and action', *International Journal of Urban and Regional Research*, 3: 309–44.

Grosfoguel, R. (1995) 'Global Logics in the Carribbean City System: the Case of Miami', in P. L. Knox and P. J. Taylor (eds.) *World Cities in a World-System*, Cambridge: Cambridge University Press: 156–70.

Keeling, D. J. (1995) 'Transport and the World City Paradigm', in P. L. Knox and P. J. Taylor (eds.) *World Cities in a World-System*, Cambridge: Cambridge University Press: 115–31.

Knox, P. L. (1998) 'Globalization and World City Formation', in S. G. E. Gravesteijn (ed.) *Timing Global Cities*, Utrecht: Netherlands Geographical Studies: 21–31.

Kunzmann, K. R. (1998) 'World City Regions in Europe: Structural Change and Future Challenges', in F.-C. Lo and Y.-M. Yeung (eds.) *Globalization and the World of Large Cities*, Tokyo: United Nations University Press: 37–75.

Malecki, E. (2002) 'The Economic Geography of the Internet's Infrastructure', *Economic Geography*, 78: 399–424.

Matsumoto, H. (2004) 'International Urban Systems and Air Passenger and Cargo Flows: some Calculations', *Journal of Air Transport Management*, 10: 241–49.

—— (2007) 'International Air Network Structures and Air Traffic Density of World Cities', *Transportation Research Part E: Logistics and Transportation Review*, 43: 269–82.

Miller, W. H. (1999) 'Airlines take to the Internet', *Industry Week*, 248: 130–34.

Nijman, J. (1997) 'Globalization to a Latin Beat: the Miami Growth Machine', *Annals of the American Academy of Political and Social Sciences*, 551: 163–76.

Rimmer, P. J. (1998) 'Transport and Telecommunications among World Cities', in: F.-C. Lo and Y.-M. Yeung (eds.) *Globalization and the World of Large Cities*, Tokyo: United Nations University Press: 433–70.

Rodrigue, J.-P., Comtois, C. and Slack, B. (2006) *The Geography of Transport Systems*, New York: Routledge.

Rozenblat, C. and Pumain, D. (2007) 'Firm linkages, innovation and the evolution of urban systems', in P. J. Taylor, B. Derudder, P. Saey and F. Witlox (eds.) *Cities in Globalization: Practices, Policies, Theories*, London: Routledge: 130–56.

Rutherford, J., Gillespie, A. and Richardson, R. (2004) 'The Territoriality of Pan-European Telecommunications Backbone Networks', *Journal of Urban Technology*, 11: 1–34.

Sassen, S. (1991) *The Global City: New York, London, Tokyo*, Princeton: Princeton University Press.

—— (2001) *The Global City: New York, London, Tokyo*, 2nd edn, Princeton: Princeton University Press.

Shepherd Business Intelligence (2005) Online. Available at www.shepsys.com. (Last accessed 19 September 2008.)

Shin, K. H. and Timberlake, M. (2000) 'World Cities in Asia: Cliques, Centrality and Connectedness', *Urban Studies*, 37: 2257–85.

Short, J. R., Kim, Y., Kuss, M. and Wells, H. (1996) 'The Dirty Little Secret of World City Research', *International Journal of Regional and Urban Research*, 20: 697–717.

Smith, D. A. and Timberlake, M. (2001) 'World City Networks and Hierarchies 1979–1999: An Empirical Analysis of Global Air Travel Links', *American Behavioral Scientist*, 44: 1656–77.

—— and —— (2002) 'Hierarchies of Dominance among World Cities: a Network Approach', in S. Sassen (ed.) *Global Networks, Linked Cities*, London: Routledge: 117–41.

Taylor, P. J. (1997) 'Hierarchical Tendencies amongst World Cities: a Global Research Proposal', *Cities*, 14: 323–32.

—— (1999) 'So-called "world cities": the evidential structure within a literature', *Environment and Planning A*, 31(11): 1901–4.

—— (2004a) *World City Network: A Global Urban Analysis*, London: Routledge.

—— (2004b) 'New Political Geographies: Global Civil Society and Global Governance through World City Networks', *Political Geography*, 24: 703–30.

Taylor, P., Catalano, G. and Walker, D. (2002) 'Exploratory analysis of the world city network', *Urban Studies*, 39: 2377–94.

Townsend, A. M. (2001a) 'Network Cities and the Global Structure of the Internet', *American Behavioral Scientist*, 44: 1697–716.

—— (2001b) 'The Internet and the Rise of the New Network Cities (1969–1999)', *Environment and Planning B*, 28: 39–58.

Zook, M. and Brunn, S. (2005) 'Hierarchies, Regions and Legacies: European Cities and Global Commercial Passenger Air Travel', *Journal of Contemporary European Studies*, 13: 203–20.

5   Airport code/spaces

Rob Kitchin and Martin Dodge

Introduction

> In our need to move, we submit to a series of invasive procedures and
> security checks that are becoming pervasive and yet are still rationalised
> through a discourse of exception – 'Only at the airport'.
>
> (Fuller and Harley 2004: 44)

Nearly all aspects of passenger air travel, from booking a ticket to checking
in, passing through security screening, buying goods in duty free, baggage-
handling, flying, air traffic control, customs and immigration checks, are now
mediated by software and multiple information systems. Airports, as we have
previously argued (Dodge and Kitchin 2004), at present consist of complex,
overlapping assemblages to varying degrees dependent to function on a myriad
of software systems, designed to smooth and increase passenger flows through
various 'contact' points in the airport (as illustrated in Figure 5.1) and to enable
pervasive surveillance to monitor potential security threats. Airport spaces –
the check-in areas, security check-points, shopping areas, departure lounges,
baggage reclaim, the immigration hall, air traffic control room, even the plane
itself – constitute coded space or code/space. Coded space is a space that uses
software in its production, but where code is not essential to its production
(code simply makes the production more efficient or productive). Code/space,
in contrast, is a space *dependent* on software for its production – without code
that space will not function as intended, with processes failing as there are no
manual alternatives (or the legacy 'fall-back' procedures are unable to handle
material flows, which means the process then fails owing to congestion).

Air travel increasingly consists of transit through code/spaces, wherein if
the code 'fails' passage is halted. For example, if the check-in computers crash
there is no other way of checking passengers in; manual check-in has been
discontinued, in part owing to new security procedures. Check-in areas then
are dependent on code to operate and without it they are simply waiting-rooms
with no hope of onward passage until the problem is resolved. In these cases,
a dyadic relationship exists between software and space (hence the slash
conjoining code/space), so that spatiality is the product of code, and code exists
in order to produce spatiality.

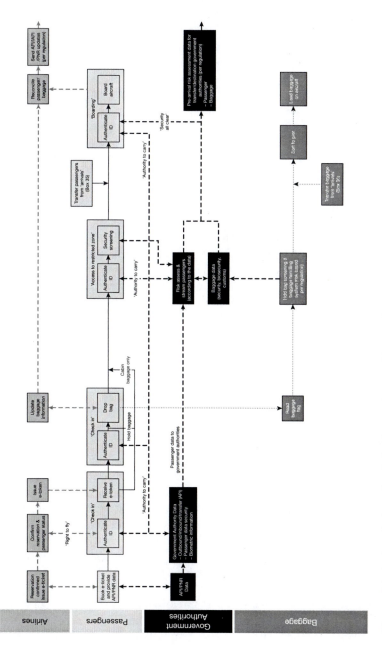

Figure 5.1 An inscription of orderly air travel created by an air travel industry expert group called 'simplifying passenger travel' to show the idealized flow of a typical departure process. The physical movement of passengers and baggage is represented by the solid lines. Many elements in the process are reliant on software and the correct exchange of digital information (represented by dashed lines in the diagram).

Source: Modified from SPT 2006: 5.

In earlier work we demonstrated how code/spaces are produced and how their operation is always contingent, relational and embodied, enacted through a process of transduction (Dodge and Kitchin 2004; 2005). Transduction is a process by which a domain shifts from one state to another, in this case from a non-coded space to a code/space. Software, we argued, alternatively modulates the production of space by altering the means by which spatialities are brought into being. It enables new, or automates old, socio-spatial processes wherein the code is essential for their deployment (see Dodge and Kitchin (2005) for a fuller explanation). Such a description might give the impression that software establishes the spatiality of much of the airport in a deterministic way (i.e. code determines in absolute, non-negotiable means everyday practices), with universal outcomes (i.e. such determinations occur in all places and at all times, in a simple cause-and-effect manner). In contrast, we would argue that the work that software does in the world is always embodied, the product of people and code. Code/space is never consistently created and experienced the same, but rather it is always produced; always in a state of becoming, emerging through individual performances and social interactions that are mediated, consciously or unconsciously, in relation to the mutual constitution of code/space (see Dodge and Kitchin 2004; 2005).

Further, we argued that code/space was most often the product of a 'collective manufacture' (Crang 1994) – of many people and systems recursively interacting with each other in multifarious ways. Indeed, airports function through multiple, interacting sets of complex socio-technological relations (see also Fuller and Harley 2004; Knox *et al.* 2005).

Code/space also varies for another reason. As we noted in Dodge and Kitchin (2004), code/spaces have accreted over time to no set master plan, with technological advances and political and economic decisions, to create interlocking assemblages. The components of these assemblages have a diverse range of owners, maintainers and licensing, accompanied by a labyrinth of contracts, leasing and service-level agreements. Further, a raft of national and international bodies and industry organizations are responsible for the setting and vetting of standards for systems where software is vital (such as aircraft navigation and air traffic control systems). As a result, the code/spaces of each airport vary in their production, the sedimentary outcome of different layers of deployments, systems, procedures and regulation laid down over years of operation. Further, code/spaces are relational, not discretely referenced to individual passengers and airports but, rather, stretched out across the whole architecture of networked infrastructure of air travel, from the locations from where tickets are initially reserved to final destination (see, for example, Bennett's (2004) detailed empirical attempt to trace some of the locations through which his personal data flowed when booking plane tickets). Code/spaces are often simultaneously local and global, grounded through the passage of people and goods, but accessible from anywhere across the network; and linked together into chains that stretch across space and time to connect start and end nodes into complex webs of interactions and transactions (e.g. ticketing and passenger name records (PNRs) held on the main global distribution systems (GDSs) (Sabre, Amadeus, Worldspan and Galileo) can be accessed from many thousands of terminals across the world). These assemblages, then, have no central control and a complexity much greater than the sum of the parts. In this sense, as we argued in Dodge and Kitchin (2004), they are an assemblage that needs to be analyzed, in Deleuze and Guatarri's (1987) terms, as striated – that is, complex, gridded, hierarchical, rule-intensive, related; and as complex systems with emergent properties (see Holland 1998; Waldrop 1994).

In this chapter we want to expand our argument that the code/spaces of air travel are emergent, relational, contingent and embodied in nature (rather than deterministic, fixed, universal and mechanistic) by demonstrating how they are brought into being through the interplay of people and code. Code, we want to demonstrate, is not law by itself (see Lessig 2000). Software's ability to do work in the world is *always* mediated by people – either through a direct interface between passenger or worker, or through gatekeepers who take the outputs of a program, interpret the results, and negotiate with a passenger(s) or fellow worker(s). What this means is that how travelers engage with software and its gatekeepers (the travel agent, check-in, security, immigration staff, and so on) and react through embodied practice varies between people and is contingent on their abilities, experiences, knowledges, and the context in which interactions occur. It is a social and cultural event, not a simple, deterministic exchange or an act of naked governmentality, and it unfolds in multifarious, ever-changing ways.

In this sense, the code/spaces of air travel are of-the-moment and performative. The airport is never repeated exactly twice and never fully predictable or ordered (though that is what systems of management and regulation aspire to). If there is a seemingly orderly pattern at a broad level it is because the various parts of the airport assemblage are citationally performed and people and systems are employed to make air travel work in particular ways. Ordering flows takes continual tuning, and as Knox *et al.* (2005: 11) note from their study of a British airport, 'the organization of "flow" is always in danger of "overflow", of disintegration into confusion and flux, where people and objects become unstuck from the smooth operation of representations and get lost in the intransigent opacity of the "mass".' Negating the occurrence of 'overflows' means the airport is remade as the airport continuously – cleaners clean; security guards patrol; food is prepared, served, cleaned away; planes land, taxi, disgorge passengers and luggage, are cleaned, refuelled, serviced, reboarded and leave; passengers and luggage flow through the various circuits and are helped on their way in various ways (by signs and flight information display screens, by printed boarding cards, by audible announcements, by customer service agents). If one spends time in the airport observing what is happening, its diverse realities become all too clear (on the sociology of airports see, for example, Gottdeiner (2001) and Pascoe (2001)). And much of this work is citationally reproduced through people and code doing work together. This becomes very apparent if a software system fails and the space fails to be produced as intended (e.g. the check-in area becomes a waiting-room), and passenger flows rupture into flux.

Airports require continuous routine maintenance, ad hoc repairs and planned renewal that are easily overlooked by passengers unless they are directly impacted (see Graham and Thrift (2007)). They exhibit 'metastability' at different scales – i.e. 'they are stable (only) in their constant instability' (Fuller and Harley 2004: 153). Given this 'collective' and 'unfinished' nature, there is always scope for 'workarounds' as airport staff in different roles adapt their interactions with software systems to cope with the pressures of on-the-ground situations; often these are 'unauthorized' actions but done with the tacit understanding of managers as necessary to circumvent systems to get the job done (e.g. sharing access accounts). There is also the ever-present potential for errors, particularly in data entry and translation within and between these software systems (see the numerous real-world stories reported on the RISKS List, catless.ncl.ac.uk/risks), while the output of software can easily be wrongly interpreted by workers and passengers (so-called 'human error'). There are also opportunities for malicious damage to vital software systems of air travel from insiders and also external attacks (e.g. computer virus damage to the US-VISIT system operated by US Customs and Border Protection agency in August 2006 caused considerable disruption; Poulsen (2006)).

In order to illustrate our arguments we draw on observant participant research as passengers.[1] This consisted of purchasing flights and traveling through a number of airports[2] between January and April 2007 and undertaking sustained observation of our own and other peoples' engagement with the software systems that are used to augment air travel. This consisted in spending time at airports in order to experience and observe the purchasing of tickets, the checking-in process, passing through security, 'hanging around' departure lounges, going to gates, boarding planes, flying, collecting baggage, passing through customs and immigration, and exiting the airport. Our observations are by no means exhaustive, but they are sufficient to add empirical weight to our argument that code/spaces unfold in diverse, negotiated and embodied ways despite the use of software designed to enforce systems of automated management (modes of governance that are automated, automatic and autonomous in nature through their use of software processing – see Dodge and Kitchin (2007)). Rather than detail examples from the full assemblage of air travel, we focus on three key sites and practices – checking in, security screening and immigration – to illustrate our argument.

Checking in

Checking in to a flight is a process that is now only achievable through software, with manual check-in discontinued for security reasons and from business logic of maintaining flow. As the Simplifying Passenger Travel group (SPT 2004: 1) states, '[t]he objective of the program is to streamline repetitive checks of passengers and their documents by collecting the information once and then sharing it electronically with all the subsequent service providers.' Increasingly the move to e-tickets also means check-in agents require 'live' data connections. While the usual procedure of queuing up to a staffed check-in desk is still commonplace, in order to save staffing costs and to increase efficiencies, airlines have been moving to self-service check-in, either at home prior to travel or through the use of self-service kiosks at airports. In all of these cases, passengers are subject to intensive, invasive surveillance and software sorting (Graham 2005) aimed at confirming identity and algorithmically assessing potential security risks, but they are also embodied, negotiated spaces and practices (see also Adey (2004); Curry (2004); Morgan and Pritchard (2005)).

Here, 'code is law' (à la Lessig) in the sense that, if the ticket or passenger is not recognizable through identification codes and personal descriptors (full name, date of birth, etc.) within the prescreening system, the passenger will initially be denied the 'right to fly', and the system might assign a passenger for extra security checks and baggage inspection while traveling. However, there is an interaction between the person(s) and code, and some problems of identity and 'trustworthiness' are negotiated through redress with agents (albeit usually with them tapping at a keyboard to correct or update systems), although the degree of negotiation is typically occluded in official evaluations of procedure and the proper working of the software system. Many of the neat boxes of idealized flow shown in Figure 5.1 are social as well as software-produced. As a result, while the experience of check-in can often be quite similar across passengers, it varies in multifarious ways as different moments of code/space are enacted through the embodied interactions of people and code.

Check-in areas traditionally consist of a row of check-in desks behind which agents sit and in front of which passengers queue. When an agent arrives at a desk to start checking in a flight, he or she first logs on to the system, accesses the flight details and sets the television monitor above the desk to reveal the destination and flight number that will be processed at that desk. As passengers reach the front of the queue, they pass over their tickets and passports/ID cards. The agent checks the ticket code against the system to confirm reservation and update the PNR, then verifies the passenger details by comparing the photo in the passport with the passenger, or alternatively, for international flights to certain destinations such as the US, scans the machine-readable part of the passport (see Figure 5.1). In US airports the ticket code and scanning will be interpreted with passenger prescreening profiling systems that will alert the check-in agent as to whether the passenger needs additional security checks further on in their passage through the airport, with this information being printed on the boarding card[3] (see GAO 2007). If the plane does not have open seating, the check-in agent will then ask about seat preference and assign seats and ask a set of predefined security questions about carry-on baggage. They then weigh (with the weight digitally displayed to the passenger) and tag the bags to go in the hold. A large label identifying the destination airport, and airports en route, along with barcodes that also identify the destination (making the bag machine-readable for systems that automatically sort and distribute bags) and owner of the bag is printed off and attached to luggage. A baggage receipt is stuck to the boarding card, and these are returned to the passenger along with the ticket, and the luggage trundles off along the conveyor belt on its own coded and tracked journey.[4]

These practices are not simply rote but are part of a social exchange between the passenger, check-in agent and information systems. Passengers ask additional questions about their travel, for example checking in to additional legs, or confirming the routing of baggage to the final destination. Check-in agents ask for additional information. And there can be frank exchanges between them when, for example, the system does not recognize the ticket or passenger, or has seemingly lost details of pre-ordered seats, or the desk is closing as a late traveler arrives, or the luggage is too heavy and the airline wants additional payment to carry it, or the flight is overbooked and the airline is seeking to hold over or re-route passengers, or the check-in agent will not check the passenger all the way through to their final destination, claiming a 'system glitch' (while the person at the next desk is having this done). We have witnessed or experienced all of these situations and others. For example, as the records from Rob's notebook document:

> The check-in agent at desk 55 types furiously on the keys and roles his eyes at the couple at the front of the queue. He batters away for another couple of minutes while the couple turn round and shrug their shoulders communicating to the rest of the queue that they are not the problem. The check-in agent, seemingly admitting defeat, picks up a phone and has a short conversation. He drops the receiver and informs the whole queue in a loud voice that they'll have to move through a parallel queue at check-in desk 54 to re-form in front of desk 53. A couple of chaotic seconds later, as people dance around each other and baggage, the queue is re-set. The agent transfers all his paperwork, tickets, baggage tape and so on, turns on the computer and taps away at the keyboard for a couple of minutes logging himself back on and accessing the right flight details. He then proceeds to book the remainder of the queue.

In this and the other examples, the situations were all solved by a combination of dialogue between people and accessing, updating and modifying information systems.

This traditional system is being supplemented with, or increasingly substituted by, self-service check-in kiosks. These consist of a touch-screen interface designed to allow the passenger to interface directly with the check-in information system, with the promise of greater efficiency for airlines and for passengers. Scott O'Leary, of Continental Airlines, claimed such kiosks meant: '[w]e are essentially queueless ... the mean check-in time is 66 seconds. For customers with no bags, it's 30 seconds' (quoted in Fishman 2004: 91). It can be argued that such kiosks, and crucially the software systems behind them, herald the next level of automated consumer service provision. The logic here, as Carr (1997; quoted in Wood 2003: 338) argues, is to produce an environment 'where security will be the only necessary human contact a passenger need make en route to the gate, freeing aeroplane employees for other activities airside.'

When approaching the kiosk, initially the passenger is asked for a booking reference code (or another form of unique identification number) and then proceeds through a set of information screens concerning security, seat selection and whether there are bags to check in. If flying to the US, the passenger is prompted to enter the machine-readable portion of their passport into a special slot and then to enter Advanced Passenger Information System (APIS; see below) details, including the address of where they will be staying. Self checking-in can be quite a prescriptive exercise in that certain fields have to be entered, but there can be options to express a degree of choice, for example with regards to seating and baggage. And just as with traditional checking-in, there is the opportunity to lie, or at least be selective with the truth with regards personal information (such as answers to: Did you pack this bag yourself? Has it been left unattended? Or in the case of APIS information, do you have a criminal record? etc.). It does not always go smoothly, and the kiosk software can crash or the network connection can freeze. It can also be quite a social experience if more than one person is checking-in at the time, or if help is needed, or if there is pressure from others to hurry up, and so on.

For example, a businessman in his forties reaches an Aer Lingus kiosk just ahead of Rob. They swap a few words of apology for nearly colliding, and Rob walks round to the next machine, which has just been vacated. He then types in his booking reference number and starts to follow the instructions. The businessman is joined by a woman and they confer about seating as he taps at the screen. After a few seconds the software on Rob's kiosk seems to have crashed. He taps at the screen, but nothing happens. He looks around for help, but all of the staff are busy helping people at other kiosks, so he heads to a free machine and starts the process again, hoping that it will work given he was already halfway through the process on a previous machine. The man and the woman are now discussing whether they are going to check one of their bags in, or whether they might get away with taking it on as carry-on luggage despite it being too large. They decide to risk it. Rob manages to complete the check-in process, changing his seat from a window at the back of the plane to an aisle near to the front. The man, woman and code were interacting with each other through a contingent, context-driven exchange.

Martin queues up to use a suite of forty self-check-in kiosks at the United Airlines terminal at Los Angeles International Airport. There is much confusion because the old way of checking in has essentially been discontinued[5] relatively recently, and many people are clearly trying to workout United Airline's 'EasyCheckin' system for the first time, including Martin.

A Mexican woman occupies the next terminal to him. She tries to use the machine but looks bewildered. She seeks to attract the attention of someone from the airline to help her, but there are only a couple of agents on the customer side of the desks for the forty machines. Afraid to try and work her way through the system in case she makes a mistake she resigns herself to a lengthy wait as she tries to get help. Like many software systems, the interface is unfamiliar to her and it takes time to adapt to, particularly in the way to respond to data entry requests and to determine how to navigate through the process. Again, this is a social, contingent, relational engagement with software.

Through security

> Effective immediately by order of the Transportation Security Administration: If you plan to travel with liquids, gels or aerosols in your carry-on bag remember 3–1–1. All liquids, gels and aerosols must be in 3 ounce or less sized containers. Containers must be placed in a 1 quart-size, clear, plastic, zip-top bag. Only one bag is permitted per traveler. It must be removed from your carry-on and placed in the security bin for X-ray screening. Remember 3–1–1 to speed your screening process.
>
> (Background PA announcement heard at US airport security check points in spring 2007, designed to order passenger behavior to smooth flow through changed screening procedures)

Like the check-in area, security checkpoints are places of queuing, boredom, chatting, fidgeting, preparation for screening, and of intense surveillance. The outward aim is ensure that no prescribed items pass through to airside and to identify and isolate passengers who might pose a security risk. It demarcates the beginnings of a sterile zone that should be devoid of proscribed people and objects. As such one might interpret the security checkpoint as a more general act of governmentality that seeks always to maintain orderly mobility. Producing this ordered, sterile space is achieved through a combination of manual observation and practices (such as uniformed staff asking questions and frisking) and automated surveillance using sensors to collect data and software to process and analyze them. The results from the sensor evaluation are interpreted and followed up by manual intervention such as bag searches. As Rob's notes illustrate:

> The bag belonging to the passenger behind is moved into operator's frame. The operator performs a set of scans. He zooms in on one section, then zooms back out again and performs the same scan routine. He then zooms back in once more and calls a colleague over. Pointing at the screen he indicates the suspected problem and they confer. The colleague then gestures to the bag's owner, a smartly dressed man in his fifties and they

head off to one side. The bag is placed on a counter and the passenger is asked some security questions and for permission to search the bag. The man concurs and all the bag's items are emptied onto a counter. The offending item is a meter long steel security cable. There is a brief negotiation, where the security official clearly sees the cable as a potential weapon and the passenger argues that it is simply for securing the laptop to a workstation. The official concedes that the man can keep the cable this time but suggests that it not be carried in carry-on luggage in future. One is very much left with the impression that not every passenger would have been allowed to keep the cable (and there is a large perspex box nearby full of confiscated items including cutlery, pen-knives, nail files, a metal ruler, a hammer, and other assorted, mostly metal, objects).

Another journey and the check-in agent is having difficulty scanning Rob's machine-readable passport. She keeps swiping it through a slot at the top of her keyboard and when it fails she checks it visually, polishes the surface clean before trying again. Eventually it seems to work, and the boarding pass is printed out. It has four Ss printed on it standing for 'Secondary Security Screening Selectee', an unassuming visible manifestation of the intensive software sorting that Rob's digital persona has been subjected to. When Rob gets to the security zone leading through to the sterilized departure area he is directed off to one side. He waits there for five minutes while periodically someone working at one of the machines calls out 'special security check'. Eventually, someone arrives, and he's moved to a new line where he is asked to remove all the usual security items. These are taken to a machine to be scanned. He is asked to stand in a GE EntryScan machine and to follow the instructions. The machine blasts air onto his clothes and hair, capturing the resultant air streams and analyzing them for explosives and narcotics using an ion trap mobility spectrometer.[6] Once the test is complete, the doors of the machine open automatically and he is then asked to sit down and wait as his bag is emptied and all of the items within are swabbed and tested for explosives. Eventually he is allowed to proceed through to the gate. As with check-in, all of these examples reveal complex social interactions between passengers, airport workers and software.

While the processes and practices are broadly similar for all passengers, they emerge in contingent and relational ways to produce divergent realities, and by no means is the code simply law. As a result, as Wood (2003: 337) notes, the security area becomes a form of theatre creating the 'Spectacle of the Frisk', a performance that 'defines contemporary air travel'. Elaborating further, Wood states:

> In a public site, we grow accustomed to viewing individuals pulled from queues, their possessions opened and studied, their bodies turned into maps of hidden threats. The gloved hands of the inspector traverse the shoulder blades, down an invisible axis toward the hips, then toward the feet. Often

accompanied by a metal detecting wand, the ordeal is mediated by questions and unspoken rules of decorum: 'Will you unbuckle your belt? May I open this bag?' Invisible lines of demarcation separate the inspected passenger from others who wait nearby. Yet all may observe the spectacle. Does he fold his underwear? What's in that zipped bag? Why did she pack so many sweaters? At once in terminal space, this interpersonal dance of touch and display reveals a network of surveillance practices that remains otherwise unnoted in public life. Beyond the local spectacle, though, we find ourselves tied within a web of individuation and deindividuation marked by perpetual surveillance.

Immigration

The gates and barriers that contain, channel, and sort populations and persons have become virtual.

(Lyon 2003: 13)

In Kitchin and Dodge (2006) we explored the use of software in regulatory and security systems designed to discipline and re-shape passenger behavior. In particular, we examined data capture and information processing systems designed not only to monitor passengers, but to build profiles to actively try and predict people who posed potential security risks. There is a shift to automatic calculation of categorical risk based on the digital body rather than individual suspicions based on the subjective assessment of the real body. To achieve this, enormous efforts and sums of money are currently being invested by governments in database systems (Figure 5.2) and new biometric identification technologies (see Amoore 2006). For example, in the US these include the US Visitor and Immigrant Status Indicator Technology (US-VISIT), APIS and Secure Flight passenger prescreening programs.

To take one of these systems: US-VISIT, which monitors all international travel in and out of the USA for the Department of Homeland Security (DHS), is being developed and operated by the Accenture-led Smart Border Alliance through a contract worth up to $10 billion (Leyden 2004). At its core, the system will consist of the integration of three existing DHS systems: the Arrival and Departure Information System (ADIS), the passenger-processing component of the Treasury Enforcement Communications System (TECS), and the Automated Biometric Identification System (IDENT) (DHS 2004). In addition to US-VISIT, passengers on international flights will continue to be prescreened by US Customs and Border Protection using APIS. APIS uses information from the machine-readable part of a passport along with PNRs supplied by air carriers (typically containing thirty-four fields of personal information) to try and identify suspect or high-risk passengers by checking for matches against a multi-agency database, the Interagency Border Inspection System (IBIS), and the FBI's National Crime Information Center wanted persons files. IBIS includes the combined databases of US Customs, US

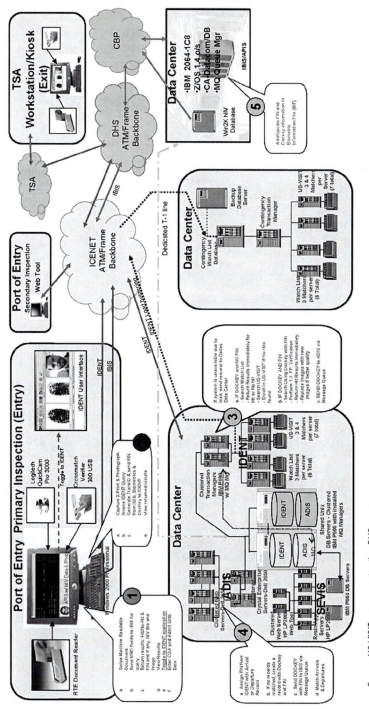

Figure 5.2 Illustrative representation of the complexity of networked information systems underlying immigration profiling by US Transportation Security Agency. This diagram is now out of date.

Source: US-VISIT procurement notice, DHS

Source: Poulsen 2006

Welcome
US-VISIT Procedures: For All International Visitors

 1: Left Index Finger
Indice Izquierdo
왼손 검지손가락
左手食指
Indicador Esquerdo

2: Right Index Finger
Indice Derecho
오른손 검지손가락
右手食指
Indicador Direito

3: Look at Camera
Mire la Cámara
카메라를 보십시오
注視相机
Olhe para a Câmera

If you have privacy concerns or questions about the safekeeping of your personal
information, please contact the US-VISIT privacy officer at usvisitprivacy@dhs.gov

 **Homeland Security**    **US-VISIT**
www.dhs.gov/us-visit

Keeping America's Doors Open and Our Nation Secure

Figure 5.3 A widely displayed poster ostensibly to inform passengers of
procedures to help maintain smooth flow through immigration control
of international passengers arriving in the US. But it can also read as a
disciplinary representation of the biometric performance necessary to
translate humans into code.

Immigration and Naturalization Service (INS), the State Department, and twenty-one other federal agencies (US Customs 2001). The data are also processed using specially designed and secret algorithms to try and predict possible high-risk passengers who do not yet appear in these databases, based on activity patterns that are deemed to diverge from normal or are out of place. In other words, a lot of data work is performed on international passengers before they have arrived in the US and as they enter the country.

It is when one passes through immigration halls that one most directly interfaces with such systems, including visual and tactile connections to code (see Figure 5.3). Here, machine-readable passports and, depending on location, biometric information are scanned, processed and interpreted, with the results outputted to gatekeepers (immigration officials). It is the gatekeeper who decides whether one gains entry to the country. Code is critical to the process, and is not easily overridden, but the decision is one made by a person interfacing with a computer and often in negotiation with the passenger, as the examples below illustrate.[7]

The man at the front of the queue has been there for a while. He is white and in his mid-forties. The immigration official keeps looking at the passport, to his computer screen, and across to the passenger. The passenger is unable to see what the official sees on his screen (a classic asymmetric power relationship of the observer and the observed, which is facilitated by the spatial configuration of the inspection booths). Talking to him afterwards, it transpires that he has recently traveled through Pakistan and then on to northern India. He has also been to Indonesia in the past twelve months and flown through Dubai. All are 'terrorist hotspots,' and he has been flagged up as a *potential* security risk. It is up to the immigration official to determine whether this potentiality might translate into a real risk while the traveler is in the US. The passenger explains that he is a professor, married to a Pakistani woman, and that he conducts research in Indonesia. The immigration official looks sceptical, and the passenger is told he will need to answer some more security questions. A few seconds later another immigration official arrives, and the passenger is escorted to an office at one side. The professor has been software-sorted (see Graham 2005) for additional attention, but the code is not simply law, as he will now have the opportunity to negotiate with the border control system. This might involve additional searches through various national and international databases or might simply consist of a conversation between official and passenger. How it proceeds will partly depend on issues such as class, nationality, profession, and so on. It is unlikely that a white, mid-forties professor with no criminal history will be classed as a security threat and he will probably be allowed into the country after a short while. A young male, non-white Muslim, however, might be a different proposition, the so-called 'flying while Arab' issue (see Fiala 2003).

The rest of the queue moves quite quickly. Then one of us is called up to the booth. The passport is swiped, fingerprints are scanned, and we are asked the reason for the visit. We state that we're attending a conference, and we have an exchange about employment and research. The official asks to see some other form of ID such as a staff card. The best that can be offered is a library and business card, which does not seem to satisfy him. Further questions follow concerning the need to attend this particular conference, how long one would be staying, where one is staying, and so on. All the time, the officer's fingers are tapping on a keyboard, and he stares at a screen, occasionally looking up. The conversation lasts a couple of minutes, and after a while it becomes apparent what is happening – the official is looking up the conference on the Internet. We volunteer the web address, and a few seconds later he seems apparently satisfied and we're allowed through. One is left wondering what would have happened if he had not found the website? Would we have followed the other professor to be asked additional questions? Would we have been allowed through? Would the questions even have arisen if we'd simply stated we were there for a holiday? Either way, software mediated the exchange in two ways – first through facilitating the search of immigration and criminal databases and second by enabling a search of the Internet.

At a different airport, a Chinese man is being berated by the immigration official. The man looks confused and scared. He holds out his passport, visa and travel documents. The immigration official pushes them away. Despite the fact that it is clear the man does not speak English he states loudly, 'I need to know where you are staying and working. I need some more documentation.' They've reached an impasse. The passenger's documents have enabled him to be software-sorted to a certain stage, but he is seemingly missing information that will complete the process. The immigration officer points to one side. 'You'll have to wait. Over there.' The Chinese man looks at the official, over at a row of seats, and back at the official. He pushes forward his documents again. 'Over there.' The official points, rolling his eyes. 'No-one speak Chinese!' he bellows patronizingly, as if this will make it easier to understand. Reluctantly the man starts to shuffle to where he is pointed, when a Chinese woman in the queue steps forward and offers to act as a translator. The official asks to see her documentation and then hesitantly agrees. Three or four minutes later the Chinese man is allowed to pass through, all smiles and bows. If the Chinese woman had not volunteered to help it is not clear how long he would have been waiting or whether he would have made it through immigration at all. Without the translator to provide the last inputs to the system, the software could not finish processing his permission to enter the country.

Another airport and we stream through immigration holding our passports open at the photo page. The official barely looks at them – he is only interested in non-EU passports – it is purely a cursory visual check. Clearly, if you are flying between EU countries then you must be a legitimate passenger – the hard work was to get into the EU in the first place. Only five people on the whole flight spend more than two seconds at the booth – three Americans and two Japanese businessmen. None of them is delayed significantly either, all clearly deemed low-risk passengers. With the exception of the five non-

EU travelers, none of the other passengers have been subject to software processing at immigration. The same happens a few weeks later at the same airport, only this time the official spends slightly longer looking at the passports of a black man and woman, and three middle-eastern-looking men. All of them have British passports, and none of them is scanned to check them in the information system. In these cases, rather than there being a negotiation between official, passenger and software, the officials have decided only to use the software to aid decision-making in a limited fashion. Usually one of the most coded of processes becomes almost entirely uncoded by manual override to one of visual inspection – an inspection that is clearly embodied and discursive, shaped by issues such as race, class and gender (as with the other examples above).

Conclusion

As these examples illustrate, airports consist of code/spaces – software purposefully mediates many of the processes and actions of air travel. However, code is not simply law – deterministic, fixed and universal. Rather, air travel emerges through the interplay between people and software in diverse, complex, relational, embodied and context-specific ways. It is an event that unfolds in multifarious, ever-changing ways. And this is not simply the case for the three parts of the assemblage we have discussed, but is also true of purchasing tickets, updating bookings, moving through and buying things in the departure lounges and gate areas, boarding the plane, the flight itself, baggage reclaim, and backgrounded systems such as building management systems, plane systems and air traffic control.

Because airports are diversely (re)produced, through the collaborative manufacture of people and code, they are certainly not the non-places as described by Augé (1995). While airports share similar architecture and processes, they are places in the same sense that small towns are, albeit with a larger throughput of people. They have diverse social relations and formations, engender meaning and attachment, represent different values and images of the locale and nation, and so on (Crang 2002). This is especially the case for the hundreds or thousands of workers and for travelers who live locally and pass through the airport regularly. And Grenoble airport with its small number of flights per day is very different to Chicago O'Hare with its thousands.

As we discuss in more detail in Dodge and Kitchin (2004), the ever greater use of software to organize, manage and produce air travel is set to grow, supported by a persuasive set of discourses that work to create a power discursive regime. These discourses include security, safety, economic rationality and increased productivity, and convenience and flexibility. Software enables securer and safer air travel by widening, deepening and automating the extent to which passengers, workers, equipment, planes and spaces are monitored and regulated through 'infallible' systems of detection and response; software enables the streamlining and automation of tasks, speeding up processes, increasing throughput, increasing efficiencies and enabling staff and product savings that can be passed on to the traveler; and software provides passengers with greater convenience and flexibility in terms of booking, itineraries of travel, passage through the airport, tracking passenger status and rewards and so on. Collectively, these discourses work to justify further investment, to make code/spaces appear as commonsense responses to particular issues, and convince travelers (and workers) of the logic of their deployment. In other words, they work to ensure that air travel will continue to consist of densely interconnected code/spaces.

Despite these efforts to further introduce ever-more deterministic forms of automated management, the code/spaces of air travel will continue to be contingent and relational in nature, the products of complex and diverse interactions between people and code. As such, we believe these interactions warrant further attention and study, requiring detailed ethnographies of aero-mobilities across peoples (passengers by class, race, gender, etc. and different kinds of worker), airports (local, national and international hubs) and countries (with differing policy, legislation and practice).

Notes

1   We acknowledge the contingent nature of our experiences and that they are only partially generalizable. While not 'elite travelers' in the conventional sense – we fly economy class – both authors undoubtedly enjoy privileged mobility as relatively affluent academics with established credit histories and being able-bodied, white males, native English-speaking and holders of EU passports. People with other identities and cultural characteristics may well have different experiences, particularly at security screening and immigration.
2   Manchester International Airport (MAN); Liverpool John Lennon Airport (LPL); Glasgow International Airport (GLA); Dublin Airport (DUB); Berlin Tegel (TXL); Grenoble Airport (GNB); Munich Airport (MUC); Chicago-O'Hare International Airport (ORD); San Francisco International Airport (SFO); Los Angeles International Airport (LAX).
3   According to some sources, there are a number of factors that will always lead to a passenger being assigned extra security checks, such as such as a one-way reservation, made within twenty-four hours. Other criteria that might lead to extra checks are: passengers traveling alone; passengers who change their flight at the last minute; passengers who pay cash for their tickets; passengers who carry no luggage; random selection. See en.wikipedia.org/wiki/Secondary_Security_Screening_Selection (accessed 25 October 2007).
4   While much of this journey is through code/space, completely dependent on software for routing, there is still scope for errors – attested to by the common passenger experience of waiting frustrated at the carousel for bags that fail to appear. For example, the UK consumer advocacy group the Air Transport Users Council statistics show that 5.6 million bags were 'mishandled' in 2006 by European-based airlines (AUC 2007). The fact that bags continue to 'lose' their human owners with such frequency shows how automation is still imperfect. Furthermore, despite intensive surveillance, employee theft from checked luggage is all too common (Heathrow airport, for example, has such a reputation for this that it has been dubbed 'Thiefrow').

5 A few conventional check-in desks exist for passengers with special needs, but the spatial organization of the terminal space and the urging of the customer service staff all 'encourage' use of the self-service kiosks. It is clear many experienced passengers find the kiosks easy to use and more efficient.

6 See GE Infrastructure Security, Entry3, www.geindustrial.com/ge-interlogix/ iontrack/prod_entryscan.html (accessed 25 October 2007).

7 Unless an entirely automated, fast-track process is used (as now operated in some airports where iris scans and biometric passports are used to verify passenger status). Even then, if there is a problem a gatekeeper will step in.

Bibliography

Adey, P. (2004) 'Secured and sorted mobilities: examples from the airport', *Surveillance and Society* 1(4): 500–19.

Amoore, L. (2006) 'Biometric borders: Governing mobilities in the war on terror', *Political Geography*, 25: 336–51.

AUC (2007) *AUC report on mishandled baggage*, London: Air Transport Users Council. Online. Available at www.caa.co.uk/docs/306/Report%20on%20mishandled %20baggage.pdf (accessed 25 October 2007).

Augé, M. (1995) *Non-places: Introduction to an Anthropology of Supermodernity,* trans. J. Howe, London: Verso.

Bennett, C. J. (2004) 'What happens when you buy an airline ticket (revisited): The collection and processing of passenger data post 9/11', paper presented at State Borders and Border Policing workshop, August 2004, Kingston, Ontario. Online. Available at web.uvic.ca/polisci/bennett/queenspaper04.doc (accessed 25 October 2007).

Crang, M. (2002) 'Between places: producing hubs, flows, and networks', *Environment and Planning A*, 34(4): 569–74.

Crang, P. (1994) 'It's showtime: On the workplace geographies of display in a restaurant in Southeast England', *Environment and Planning D: Society and Space*, 12: 675–704.

Curry, M. (2004) 'The profiler's questions and the treacherous traveler: Narratives of belonging in commercial aviation', *Surveillance & Society*, 1(4): 475–99.

Deleuze, G. and Guattari, F. (1987) *A Thousand Plateaus: Capitalism and Schizophrenia*, trans. B. Massumi, Minneapolis, Minnesota: University of Minnesota Press.

DHS (2004) *US-VISIT Program, Increment 2, Privacy Impact Assessment*, Washington DC: Department of Homeland Security. Online. Available at www.dhs.gov/interweb/ assetlibrary/US-VISIT_PIA_09142004.pdf (accessed 25 October 2007).

Dodge, M. and Kitchin, R. (2004) 'Flying through code/space: The real virtuality of air travel', *Environment and Planning A*, 36(2): 195–211.

—— and —— (2005) 'Code and the transduction of space', *Annals of the Association of American Geographers*, 95(1): 162–80.

—— and —— (2007) 'The automatic management of drivers and driving spaces', *Geoforum*, 38(2): 264–75.

Fiala, I. J. (2003) 'Anything new? The racial profiling of terrorists', *Criminal Justice Studies*, 16(1): 53–8.

Fishman, C. (2004) 'The toll of a new machine', *FastCompany*, May: 91. Online. Available at www.fastcompany.com/magazine/82/kinetics.html (accessed 25 October 2007).

Fuller, G. and Harley, R. (2004) *Aviopolis: A Book About Airports,* London: Blackdog Press.

GAO (2007) *Aviation Security: Efforts to strengthen international passenger prescreening are under way, but planning and implementation issues remain,* Washington DC: United States Government Accountability Office. Online. Available at www.gao.gov/cgi-bin/getrpt?GAO-07–346 (accessed 25 October 2007).

Gottdeiner, M. (2001) *Life in the Air: Surviving the New Culture of Air Travel,* Oxford: Rowman and Littlefield.

Graham, S. (2005) 'Software-sorted geographies', *Progress in Human Geography* 29(5): 562–80.

Graham, S. and Thrift, N. (2007) 'Out of order: Understanding repair and maintenance', *Theory, Culture & Society,* 24(3): 1–25.

Holland, J. H. (1998) *Emergence: From Chaos to Order,* Oxford: Oxford University Press.

Kitchin, R. and Dodge, M. (2006) 'Software and the mundane management of air travel', *First Monday,* 11(9). Online. Available at firstmonday.org/issues/special11_9/kitchin/ (accessed 25 October 2007).

Knox, H., O'Doherty, D., Vurdubakis, T. and Westrup, C. (2005) 'Enacting airports: Space, movement and modes of ordering', Evolution of Business Knowledge (EBK) Working Paper, 2005/20. Online. Available at www.ebkresearch.org/downloads/workingpapers/wp0520_knox_etal.pdf (accessed 25 October 2007).

Lessig, L. (2000) *Code and Other Laws of Cyberspace,* New York: Basic Books.

Leyden, J. (2004) 'Accenture wins $10bn homeland security gig', *The Register,* 2 June. Online. Available at www.theregister.co.uk/2004/06/02/accenture_homeland_security_win/ (accessed 25 October 2007).

Lyon, D. (2003) 'Surveillance as social sorting: computer codes and mobile bodies', in D. Lyon (ed.) *Surveillance as Social Sorting: Privacy, Risk and Digital Discrimination,* London: Routledge.

Morgan, N. and Pritchard, A. (2005) 'Security and social 'sorting': Traversing the surveillance-tourism dialectic', *Tourist Studies,* 5(2): 115–32.

Pascoe, D. (2001) *Airspaces,* London: Reaktion.

Poulsen, K. (2006) 'Border security system left open', *Wired News,* 4 December. Online. Available at www.wired.com/science/discoveries/news/2006/04/70642 (accessed 25 October 2007).

SPT (2004) *Air Passenger Process.* Simplifying Passenger Travel (SPT). Online. Available at www.spt.aero/files/downloads/21/SPT_Air_Passenger_Process_June_2004.pdf (accessed 25 October 2007).

—— (2006) *Ideal Process Flow V2.0 report,* Simplifying Passenger Travel (SPT), 1 December. Online. Available at www.spt.aero/files/downloads/21/IPF_V20_30_Nov_06.pdf (accessed 25 October 2007).

US Customs (2001) 'New Law Makes APIS A Must For International Air Carriers', *US Customs Today,* December. Online. Available at www.cbp.gov/xp/Customs Today/2001/December/custoday_apis.xml (accessed 25 October 2007).

Waldrop, M. M. (1994) *Complexity: The Emerging Science at the Edge of Order and Chaos,* Harmondsworth, Middlesex: Penguin Books.

Wood, A. (2003) 'A rhetoric of ubiquity: Terminal space as omnitopia', *Communication Theory,* 13(3): 324–44.

6   Air craft

Producing UK airspace

Lucy Budd

It's a tricky business directing traffic at 35,000 feet. There are no traffic lights. No road signs. No roundabouts either. But, fortunately, for the two million flights and 220m passengers that pass though UK airspace every year, there's NATS . . .

(NATS recruitment advertisement)

The United Kingdom contains some of the most densely trafficked airspace in the world. In 2006, an average of over 5,400 commercial flights a day shared the skies with hundreds of military jets, private aircraft, helicopters, airships, hot-air balloons and gliders. They were protected from collision by the skill and vigilance of their pilots and air traffic controllers, the careful arrangement of airways and control zones, and increasingly sophisticated collision avoidance software, yet the only time many of us get to hear about this complex, largely invisible, interlocking aerial geography of command and control is when things go wrong and flights are delayed, diverted or cancelled owing to adverse weather conditions, computer failure or industrial action. Most of the time, the safe, efficient and punctual production of airspace forms a vital part of a largely taken-for-granted airworld.

While much has been written about the development and utilization of new aeronautical technologies, the evolution of airline networks, the growth of airports and aviation's apparent ability to 'shrink' global space-time, airspace remains an under-researched and under-theorized site of aeronautical activity. Where it has been considered, it has often been described as a mere 'conduit' or 'space of flows', negating any detailed investigation into how it is socially produced, maintained and contested through ongoing practices of management, negotiation and opposition. Such is the paucity of research into the everyday, yet largely hidden, spatial practices of Air Traffic Control (ATC) and the piloting of commercial aircraft, those not directly involved in its production are largely ignorant as to how airspace 'works' and why the sky is configured and used in particular ways. This has important implications at a time of continued passenger demand and widespread public opposition to airport expansion.

In order to bring questions of airspace production to the forefront of academic enquiry, this chapter contains five distinct, but intrinsically inter-related, sections. The first, entitled 'crafting the sky', provides an overview of the development of airspace legislation from the early twentieth century to the present day. It examines how the sky has been crafted into an important geopolitical space that is simultaneously governed by a multitude of domestic and international laws. In an effort to expose the complex 'hidden' geographies of the air that have been created, section two provides a brief description of the contemporary structure and classification of UK airspace. Sections three and four then explore how these unique aerial spatialities are reproduced and mediated by practices of ATC and the piloting of commercial aircraft. The final section draws on recent examples of anti-airport protest in the UK to suggest that airspace is not only produced 'in the air' by air traffic controllers and pilots, but is actively negotiated and contested on the ground by communities who oppose its use.

Crafting the sky

The development and utilization of powered flight in the early twentieth century demanded the extension of traditional Cartesian understandings of territory to embrace the third (aerial) dimension. While some praised the freedom and emancipation flight afforded, predicting it would bind the nations of the world together in a new era of international peace and understanding (Finch 1938), others were concerned about the combined military and com-mercial threat aircraft posed. As Dargon (1919: 146) noted, 'whereas other vehicles ... are compelled to keep to existing tracks, aircraft are free to manoeuvre in space and can rapidly and easily surmount all obstacles which have hitherto constituted effective barriers to other forms of locomotion', and individual states felt compelled to defend themselves against uninvited or hostile 'winged visitors' through a collection of hastily formulated aerial legislation (Brittin and Watson 1972).

As long as a pilot took off, flew within a state's navigable airspace and landed within its national borders there was no problem, but the challenge international services posed to the territorial integrity of individual states produced one of the longest and most acrimonious debates in aeronautical politics. Nation-states sought to cede as little and seize control of as much airspace as possible and manipulated international agreements governing economic regulation for their own commercial advantage while retaining control over their borders for reasons of defence and national security (Petzinger 1995).

Countries with rapidly developing aviation interests, including the UK and US, advocated complete freedom of the skies, cautioning against any bureau-cratic intervention (other than that which helped secure their aerial hegemony), arguing,

The road of the air is a free and universal thoroughfare for all mankind. As wide as the world, and almost everywhere navigable, it is unhampered by any barrier, obstacle or limitation . . . Any restriction to its usage will be an arbitrary restriction imposed by the will of man.

(Burney 1929: 167)

One of the main obstacles to agreement was that, while national claims to land, lakes, rivers and adjoining seas had been common since Roman times, claims to airspace were entirely new concepts. Nevertheless, it was agreed that some form of transnational regulation was required, and the first coherent attempt to bring international air services under unified control occurred in Paris in 1910. However, the mutually incompatible visions held by the representatives of different aerial nations meant that unanimous agreement on the use and regulation of airspace was not forthcoming (Veale 1945).

Following the first scheduled passenger flight between England and France in August 1919, the production and control of global airspace became a matter of intense political concern and, as the twentieth century progressed, the sky was parcelled out between nations and subdivided into a number of discrete 'blocks' that were subject to different rules and regulations. A plethora of bilateral and multilateral air service agreements were signed that stipulated which airlines could fly, which airports (and hence airspace) they could use, how frequently the services could operate, and the airfares that could be charged (see Millichap 2000). European flag-carriers, including British Airways, Iberia and Lufthansa, thus operated in a highly protected market, insulated from any form of effective competition. It was not until the late 1980s that any change occurred. Increased public dissatisfaction with high airfares, combined with the rise of neo-liberal economic ideologies and pressures on public spending, encouraged European governments to embark on an ambitious programme of air transport liberalisation (Balfour 1994).

The removal of anti-competitive legislation, through three progressive packages of liberalization measures in the 1990s, revolutionized the industry and allowed new airlines to enter the marketplace for the first time. Many chose to undercut the airfares charged by traditional carriers by eschewing traditional in-flight 'frills' and operating frequent short-haul flights between secondary, less congested, regional airports (Calder 2002; Lawton 2002). Lower fares stimulated unprecedented passenger demand and a dramatic rise in passenger numbers, but the resulting increase in flights, particularly at smaller airports, posed a number of challenges for air traffic control. As one senior controller commented,

the skies are now full of Ryanairs and easyJets going to places you've never heard of. You suddenly find you've got 25 aircraft all wanting to go (from the UK) to Malaga at 7 a.m. on a Saturday morning and there simply isn't room.[1]

Another remarked,

> often the first I know about a new route is when I see an advert for it on
> a bus shelter. Airlines and passengers just assume they can fly wherever
> and whenever they want, but the reality is rather different. You might think
> the sky is limitless, but believe me, it isn't.[2]

Indeed, the oft-vaunted 'freedom' of the air is largely an illusion, and the space available for different types of flight is restricted. The existing airspace structure requires commercial aircraft to fly along strictly defined airways (the equivalent of aerial roads in the sky) and circumnavigate large areas of sky that are reserved for military use. The UK's geographical site and situation between the old and new worlds also mean that up to 80 per cent of the capacity of certain airspace sectors can be occupied by aircraft flying between North America and Continental Europe, leaving little room for domestic or intra-European flights.

Today, the provision, regulation and use of UK airspace are becoming increasingly politicized. Government and industry regulators want a safe, competitive and efficient airspace system. Airlines crave the freedom and flexibility to fly where they want, when they want to, as cheaply as possible. Military and general aviation users require access to airspace for training and recreation purposes, and environmental groups and airport communities complain about levels of aircraft noise and pollution and seek to restrict the industry's growth.

Ordering the sky

At a national level, UK airspace is governed and administered by the Civil Aviation Authority (CAA) and NATS, the part-privatized national air traffic services provider, in accordance with domestic and international law. All flights within the UK's 350,000 square miles of sovereign airspace are conducted according to one of two rules of flight – visual flight rules (VFRs) or instrument flight rules (IFRs) – which determine where and when pilots can fly. Under VFR protocol, pilots assume complete responsibility for aerial navigation and the safe conduct of their flight. Newly qualified pilots are only permitted to fly in good weather and good visibility during daylight hours (though experience and the acquisition of additional licence ratings may modify these conditions). IFRs, in comparison, allow suitably qualified pilots to fly in controlled airspace (upon receipt of ATC clearance), 24 hours a day in virtually all weathers. The majority of commercial flights in UK airspace are flown according to IFR, while most general aviation users operate under VFR conditions.

To help manage the diverse operational requirements of different airspace users, UK airspace is divided into two geographical regions. 'London' is administered from the en-route air traffic control centre at Swanwick,

Hampshire, while 'Scottish' sectors are controlled from Prestwick. Both regions are divided vertically, with a Flight Information Region (which is active from the ground to 19,500ft) and an Upper Flight Information Region (for airspace above 19,500ft) in each. Different sections of airspace within these regions are further classified as being 'controlled' or 'uncontrolled', depending on the nature and volume of traffic flowing through them. Controlled airspace (i.e. that which falls under the jurisdiction of ATC) can take many forms, from en-route high-altitude airways to local airport control zones, while uncontrolled airspace is relatively 'free' and can be accessed by anyone with a valid licence. To identify the different types of airspace and determine the rules that apply in each, each sector is designated as one of seven 'classes' (identified by the letters A-G), where Class A is subject to the most control, and Class G the least.[3] These designations create a highly complex web of different control zones and sectors, all of which are effective between different altitudes, subject to different rules and regulations, and may only be active for certain periods of time. Knowing where you are, and when and where you may fly, is thus crucial to the maintenance and safe production of airspace.

To compound the complexity, some areas of sky are permanently off-limits to civilian aircraft for reasons of safety and/or national security. These restricted areas include military training zones, areas around certain power installations and defence establishments, and certain wildlife reserves. Temporary restricted areas may also be introduced during major sports events or airshows. During the UK stage of the Tour de France in July 2007, six temporary restricted areas were activated above parts of London and the southeast to protect the television helicopters and other aircraft monitoring the race. Temporary restricted areas may also be established around airshows to protect both the performers and other airspace users. Details of the lateral, vertical and temporal extent of these restrictions are communicated through airspace charts, Notices to Airmen (NOTAMs) and pre-flight bulletins. As there are no fences or 'keep out' notices in the sky, the onus is on the pilot (and, to a lesser extent, the air traffic controller) to ensure the boundaries of different types of airspace are not violated. However, the system is not infallible, and controlled airspace can be, and often is, encroached by unauthorized aircraft. In 2006, 633 separate airspace infringements were reported to the CAA. Though the majority did not pose a collision risk, a small number resulted in serious 'airprox' events (so-called 'near misses'). Fortunately, none of these incidents resulted in a mid-air collision, but it is estimated that just one infringement can affect up to thirty other aircraft, delay as many as 5,000 passengers, and cost over £50,000 in wasted fuel (CAA 2007).

In the early years of passenger flight, pilots navigated with reference to major landmarks such as roads and railway lines, but, as the network of passenger services grew throughout the 1920s, identification codes were painted on top of railway stations, barns and hangars to help pilots determine their exact location from the air. This system, however, required aircraft to remain below the cloud-base and converge at a few key navigation points, and simultaneously condemned passengers to an uncomfortable ride and increased the risk of mid-air collision. In 1922, seven people died in a mid-air collision over northern France, and a decision was taken to regulate air routes across the English Channel. As a precursor of the modern airway system, pilots flying between London and Paris were instructed to remain east of Ashford, Etaples and Ecouen when flying towards the French capital, and west of them on their return. To aid compliance and assist with navigation, radiotelephony stations were constructed to enable ground controllers to communicate with pilots over the Channel (National Air Traffic Services 2005).

By the 1930s, rising numbers of aircraft necessitated the creation of specific arrival and departure routes at airports to ensure aircraft remained a safe distance apart. The principles of this system form the basis of current airport operations, with inbound and outbound aircraft following predetermined arrival and departure routes. Current standard arrival routes (STARs) and standard instrument departure routes (SIDs) are designed to ensure aircraft can leave and join the en-route airways safely and efficiently. At large airports, these routes are highly complex, and the specific procedures pilots must follow are communicated through specialist charts (see Figures 6.1 and 6.2).

As the twentieth century progressed, a national network of very high frequency omnidirectional range (VOR) radio beacons was established to aid aerial navigation and define the dimensions and contours of UK airspace. VOR beacons transmit a coded signal on a specific radio frequency that enables aircraft to 'home in' on them from any direction and 'turn corners' at the intersection of two or more beams. Individual beacons are identified by a name and a three-letter abbreviation which, like the airspace sectors above them, often have some basis in 'real world' geography, such as 'Clacton' (CLN) on the Essex coast, 'Brookman's Park' (BPK) near the famous motor racing circuit, and 'Trent' (TNT) in the Peak District.

To help controllers and pilots monitor a flight's progress, over 820 reporting points and/or waypoints are located along the airways. Some of these 'Name Code Designators' also reflect their geographical location, for example 'RUGBY' and 'LESTA' (Leicester) in the English Midlands, 'MIRSI' (as in River Mersey) north of Liverpool, and 'KIDLI' near Oxford Kidlington airport. Some may also contain an implicit 'local' connection, such as 'ABBOT' near Stansted airport (named after a local Essex beer) and 'UPDUK' in Leicestershire (which has been linked to the local colloquial greeting 'hey up me duck'), but as traffic volumes have grown, and additional routes have been introduced, new names have emerged that bear no relationship to ground-based features below. Some are named after British flora and fauna (examples include 'HAZEL', 'BUZAD' and 'FINCH'), or female names (including 'KELLY', 'LINDA' and 'KATHY'), while a group of waypoints over the English Channel are named after famous nautical heroes. While it is claimed that software alone determines waypoint names, a degree of humour apparently creeps in – with 'RUGID' over the Scottish highlands, 'BARMI' over the North Sea, 'NEDUL' and 'THRED' near the South Coast, and 'GINIS'

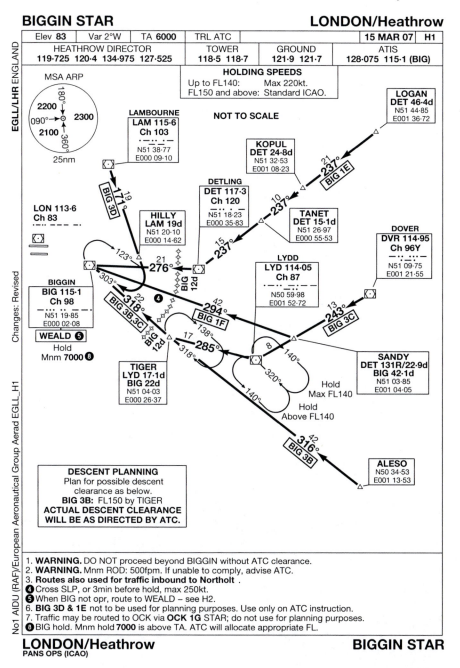

BIGGIN STAR   **LONDON/Heathrow**

| Elev **83** | Var 2°W | TA **6000** | TRL ATC | | | **15 MAR 07** | **H1** |

| HEATHROW DIRECTOR 119·725 120·4 134·975 127·525 | TOWER 118·5 118·7 | GROUND 121·9 121·7 | ATIS 128·075 115·1 (BIG) |

EGLL/LHR ENGLAND

Changes: Revised

No1 AIDU (RAF)/European Aeronautical Group Aerad EGLL_H1

MSA ARP

2200 / 180° / 2300 / 090° / 2100 / 360° / 25nm

HOLDING SPEEDS
Up to FL140: Max 220kt.
FL150 and above: Standard ICAO.

NOT TO SCALE

LAMBOURNE **LAM 115·6 Ch 103** N51 38·77 E000 09·10

LON 113·6 Ch 83

HILLY LAM 19d N51 20·10 E000 14·62

BIGGIN **BIG 115·1 Ch 98** N51 19·85 E000 02·08

WEALD ❺ Hold Mnm **7000 ❽**

LOGAN **DET 46·4d** N51 44·85 E001 36·72

KOPUL **DET 24·8d** N51 32·53 E001 08·23

DETLING **DET 117·3 Ch 120** N51 18·23 E000 35·83

TANET **DET 15·1d** N51 26·97 E000 55·53

DOVER **DVR 114·95 Ch 96Y** N51 09·75 E001 21·55

LYDD **LYD 114·05 Ch 87** N50 59·98 E001 52·72

BIG 1E 237° 21
BIG 3D 171° 19
BIG 12d 276° 21
123°
303° 22
BIG 3B 3C 318°
BIG 12d 318°
BIG 1F 294° 42
285° 17
237° 15
237° 10
237°
243° 13 BIG 3C
❹
138°
140°

SANDY **DET 131R/22·9d BIG 42·1d** N51 03·85 E001 04·05

Hold Max FL140
Hold Above FL140
320°
140°
8

TIGER **LYD 17·1d BIG 22d** N51 04·03 E000 26·37

316° 42 BIG 3B

ALESO N50 34·53 E001 13·53

DESCENT PLANNING
Plan for possible descent clearance as below.
BIG 3B: FL150 by TIGER
ACTUAL DESCENT CLEARANCE WILL BE AS DIRECTED BY ATC.

1. **WARNING.** DO NOT proceed beyond BIGGIN without ATC clearance.
2. **WARNING.** Mnm ROD: 500fpm. If unable to comply, advise ATC.
3. **Routes also used for traffic inbound to Northolt** .
❹ Cross SLP, or 3min before hold, max 250kt.
❺ When BIG not opr, route to WEALD – see H2.
6. **BIG 3D & 1E** not to be used for planning purposes.
7. Traffic may be routed to OCK via **OCK 1G** STAR; do not use for planning purposes.
❽ BIG hold. Mnm hold **7000** is above TA. ATC will allocate appropriate FL.

LONDON/Heathrow   **BIGGIN STAR**
PANS OPS (ICAO)

Figure 6.1 The BIGGIN STAR into London Heathrow. Note the distinctive coded graphic depiction of the bearings, beacons and waypoints delimiting the various routes and the textual description of the procedures that are to be followed (courtesy of European Aeronautical Group).

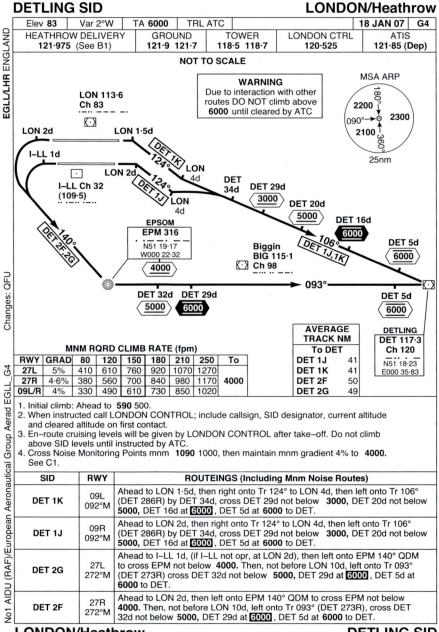

Figure 6.2 Depiction of the standard instrument departure route from London
Heathrow via DETLING (courtesy European Aeronautical Group).

over the Irish Sea. Unlike beacons, waypoints are not marked by any built infrastructure and are 'invisible' markers designed to regulate and control flows of aircraft. Many waypoints only exist in upper or lower airspace, creating an invisible vertical geography of striated layers of airlanes and airways whose positions are irrelevant for those aircraft not operating between those altitudes.

Like roads, airways are classified and given alphanumeric identifiers, and the route a flight will follow is detailed on the flight plan as a string of letters and numbers. For example 'DTY-A47-WOD-BIG-UL9-DVR' describes a route from the Daventry beacon (DTY) to Dover (DVR) via airway A47, the

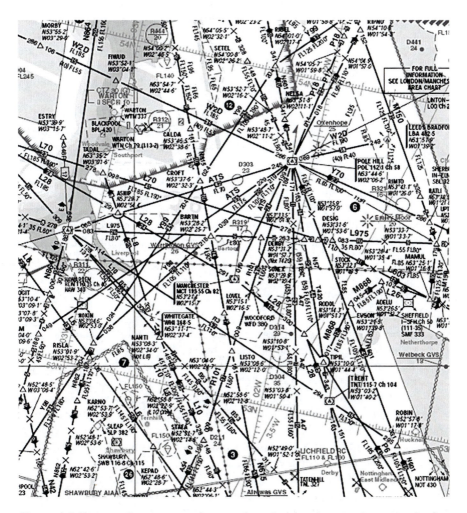

Figure 6.3 Extract of an en-route airspace chart showing the complex network of airways and reporting points over north-west England (courtesy European Aeronautical Group).

beacons at Woodley (WOD) and Biggin Hill (BIG), and airway Upper Lima Nine (UL9). To ensure individual airspace sectors are not overloaded, flow management computers at Eurocontrol in Brussels analyze the spatial and temporal profile of all flights that are planning to use European airspace for some or all of their journey and impose slot restrictions or issue alternative routings, where necessary, to smooth out the flow.

To help pilots comprehend increasingly complex air routes and airspace sectors, dedicated aeronautical charts began to be published to aid navigation and spatial orientation. The first series of aerial navigation charts designed specifically for commercial use appeared in the immediate aftermath of World War II and were produced to facilitate the development of regular international passenger services. At one level, early airspaces were relatively easy to map, and the physical architecture and topology of the network were simply superimposed on a conventional map and represented using an appropriate form of cartographic visualization. But as aircraft began flying progressively further, faster, longer and higher, new universal classification systems had to be devised.

The current portfolio of paper and electronic aerial navigation charts available to pilots is extensive and includes everything from small-scale high- and low-altitude en-route IFR charts, regional airspace information supplements, and aerodrome booklets that show the layout of runways, taxiways and gates and detail the specific arrival and departure procedures that should be followed at each facility, to VFR charts for the private pilot. All these publications code the sky in different ways and require the user to be familiar with the distinct language and symbology of airways, airspace exclusion zones, minimum safe altitudes, radar vectoring areas, and associated information (see Figure 6.3).

Controlling the sky

Irrespective of the number of maps and charts depicting where different types of aeromobility can occur, aircraft cannot enter controlled airspace without ATC authorization, and the day-to-day production of UK airspace relies on controllers producing space for aircraft. In common with the rest of the industry, the spatial practice of ATC is highly regulated and is mediated by specialized technology that helps controllers 'see' the airspace under their command and enables them to order and police the sky at a variety of scales, often from remote sites.

Radar is one of the most important tools of ATC and is employed at all control centres to monitor the progress of individual flights and help controllers visualize traffic flows. Modern radar involves two discrete systems operating in tandem, primary surveillance radar (PSR) and secondary surveillance radar (SSR). PSR produces coloured 'blips' showing the location of any object (including aircraft, high-sided vehicles, storm clouds and areas of high ground) that causes an echo to be reflected back to the receiver. In order to positively distinguish aircraft from other 'ghost' echoes, all passenger aircraft over a certain weight are required by law to carry a small radio device, called a transponder, in the flightdeck. Transponders automatically respond to interrogation from ground-based radar pulses and send a unique coded identification 'squawk' signal, which contains salient information about the aircraft's speed, altitude and rate of climb or descent, back to the ground. ATC computers then translate these transponder signatures back into flight data, providing controllers with information about the operator, call-sign, altitude, origin/destination, speed and rate of climb or descent (if it exceeds 500 feet per minute) of individual aircraft, before showing these data alongside the relevant 'blip' depicting the aircraft's physical position in space.

The resulting two-dimensional images of aircraft flying through three-dimensional space are layered on top of a static grid of lines and symbols demarcating different airspace sectors and showing the position of airports, navigation beacons and waypoints, and produce a complex assemblage of different objects flying through multiply encoded spaces. The responsibility for interpreting these images rests with individual controllers, and many report they develop detailed three- and four-dimensional mental pictures of the airspace they are working. One controller commented that aircraft 'bounce off' flight levels in her head, while a colleague remarked that the construction of mental images 'is not something we do consciously, it just happens – I look at a radar display and instinctively see it in 3D'.[4]

To lessen the risk of incomprehension and misunderstanding, all ATC messages are conducted in English, and each sector of airspace is administered using a dedicated airband frequency to minimize interference. As every pilot can hear all the transmissions between the controller and the other aircraft operating on that particular frequency, they can develop situated under-standings of the relative position and trajectory of the air traffic around them. The use of read-back, whereby flight crew repeat the controller's instruction alongside their call-sign, acts as another safety device, ensuring all instructions have been received and understood. Nevertheless, in 2004–2005, 538 separate communication incidents were reported in UK airspace that involved pilots either mishearing or misunderstanding ATC instructions (Jones 2005).

To help controllers keep track of the clearances and instructions they issue, they continually annotate flight progress strips (FPSs), which accompany every flight throughout its journey. Before take-off, data about a particular service (including call-sign, operator, aircraft type, intended routing, requested altitude, anticipated airspeed, scheduled time of arrival or departure, and details of any en-route delays) are uploaded from the flight plan, encoded and printed onto lengths of card (or displayed on electronic screens in some control centres). Once processed and approved, flight data are automatically sent to all the control centres that will handle that flight. Before an aircraft departs or arrives in a particular sector of airspace, these strips are printed, placed in coloured holders to differentiate between different types of service, and positioned in chronological order in strip-racks near the radar screen. New strips are inserted at the top of the rack furthest from the controller and, as flights land or leave the sector, the remaining strips move down to take their place, bringing aircraft closer to the controller in time and space (with the relative 'height' of strips on the rack standing proxy for either the altitude of arriving aircraft as they descend towards the airport or the sequence in which they must be handled).

Once a strip becomes live and the aircraft to which it refers is under active control, every salient detail about the flight, including heading changes, altitude clearances, speed restrictions or special instructions, is added to update the basic printed information. As these instructions are dependent on emerging contingencies, no two strips are ever the same, and individual controllers literally 'author' the sky to reflect their personal view of the airspace under their command. Depending on traffic volumes and weather conditions, individual strips can become covered in annotations, showing how the process of control creates airspace in flexible ways. This act of inscribing information defines the airspace in the controller's own terms, but while every controller 'produces the sky' in different ways, the information is presented in a universally structured manner (Figure 6.4).

While the spatial practice of air traffic control appears very prescribed, with tasks mediated by international regulations, manuals and protocols, controllers do have the flexibility to choreograph the production of airspace in different ways according to emerging contingencies. The importance of controller discretion, or flexibility, within defined operating parameters should not, therefore, be underestimated. A violent thunderstorm may require aircraft to deviate from prescribed routes, or an in-flight emergency may necessitate prioritizing one aircraft above all others. However, any disruption to normal flow patterns, no matter how seemingly slight, can have significant knock-on effects on the whole network, with delays in one sector affecting traffic hundreds of miles away. Controllers thus seek to keep aircraft moving through the sky as safely and efficiently as possible, but rely on pilots to enact their instructions. The role of the human pilot is therefore also of fundamental importance to the production of airspace and must be explored.

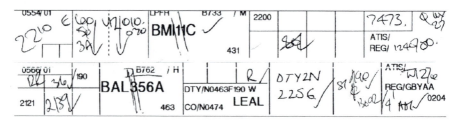

Figure 6.4 Annotated flight progress strips detailing a flight from Faro (top) and a departure to Alicante (bottom).

Navigating the sky

Flying a commercial aircraft is an inherently spatial act, where the interdependence and interaction between multiple encoded infrastructures, technologies and practitioners are integral to the production of airspace. Yet, far too often, social scientists have treated aircraft as objects to be observed, their routes plotted and their service frequencies analyzed, while the everyday practices of piloting that produce airspace for thousands of passengers every day have been largely ignored.

As far as many passengers are concerned, catching a flight is relatively straightforward: you book and pay for your ticket, present yourself at the correct airport on the right day in time for your flight, with your luggage and identification, check in, clear security, board the appropriate aircraft, and deplane several hours later at your destination. Indeed, some industry trade names, including Air*bus*, *easy*Jet, and the now-defunct Sky*train* (my emphasis), encourage the notion that flying is a routine, everyday activity. However, before the passengers check in, tens of thousands of electronic transmissions and dozens of pieces of paperwork will have been produced, circulated and checked to ensure that the right aircraft is at the correct gate at the right airport at the right time, fully serviced, fuelled and crewed (see Peters (Chapter 8), in this book). All of these documents, from load sheets and flight-plans to weather reports and checklists, combine to produce airspace in a particular way for a particular flight.

While some scholars have begun investigating the complex relationship that exists between the introduction of new forms of technology and the production of certain types of social space (see Graham and Marvin 2001; Thrift and French 2002), little or no research has been conducted into how commercial airline pilots develop and communicate situated understandings of airspace through the interpretation of flightdeck displays and the routine practice of completing flight-phase-related activities. This academic lacuna is due, in part, to strict security protocols that render permission to conduct such research problematic, and because aviation's technical language and unique operating procedures render it an intimidating prospect for study. While understandable, this omission is serious, as many tragic accidents involving commercial aircraft have been attributed to pilots exhibiting poor spatial awareness or misinterpreting or unquestioningly trusting malfunctioning flightdeck instruments (Beaty 1991; Faith 1996). Indeed, it could be argued that rarely is the accurate production and unambiguous interpretation of space, as mediated through increasingly sophisticated electronic avionics software, more critical than on the flightdeck of a commercial airliner. Building on a range of sociological research into the mundane and practical elements of work, social interaction, and technology in complex organizational environments (see Heath *et al.* 1999), this section examines the work of flightcrew, who continually interact with complex technology to pilot their aircraft through the sky in accordance with a highly regulated ATC system.

In recent years, geographers have begun to explore the extent to which computer software (or code) is deeply embedded within the infrastructure of contemporary capitalist societies and how it has become central to the spatial formation of everyday life (Dodge and Kitchin 2004; 2005; Graham 2005). In the context of commercial aviation, the sheer number of computer components installed in modern aircraft reveals the extent to which computer code mediates the production of airspace. The Boeing 777 is controlled by over 2.6 million lines of software code (Norris and Wagner 1996), while the new A380 'super jumbo' contains over 350 miles of wiring (Fortson 2007).

Dodge and Kitchin (2004) have suggested that the increasing sophistication of electronic aircraft systems means pilots fly through real space virtually, using a plethora of digital instruments and sensors. Drawing inspiration from the work of Castells (1996), they explore how the production of specialist computer code mediates the production of different 'code/spaces' of aviation, from check-in counters, security checkpoints, departure lounges and aircraft cabins, to baggage reclaim belts and retail areas. They posit that the use of computerized systems at every stage of every flight means the practice of travelling by air has become virtualized to the extent that corporeal aero-mobilities are totally reliant on the safe, efficient and routine functioning of a multitude of different networked computer systems, from reservation databases to flight-planning software and passenger manifests. For the most part, these systems are taken for granted, and dependence on them is only exposed when a computer breakdown grounds flights or a malfunctioning baggage system misroutes luggage. According to their thesis, modern aircraft can be considered contemporary 'code/spaces' *par excellence* on account of the number of, and near total reliance on, sophisticated avionics and life-support systems (see Kitchin and Dodge (Chapter 5), elsewhere in this book).

Given the inherent complexity of aircraft systems, and the need to monitor their performance, modern flightdecks feature a seemingly bewildering array of buttons, dials, levers, lights and electronic displays, all of which convey information about different aspects of the aircraft's operation and performance. These instruments are grouped according to function, and pilots are trained to check them in a particular order and consider a flight as a series of flows of information.

The primary flight displays are situated immediately in front of both pilots and convey all the 'basic' information about the flight, including the aircraft's attitude relative to the horizon, altitude, airspeed and vertical speed. Neighbouring navigation displays present information on the aircraft's track and routing, as well as information from the weather radar and the onboard collision avoidance software. The latter system, Traffic Control and Collision Avoidance System (TCAS), enables pilots to 'see' the position of other air traffic in the vicinity by providing abstract two-dimensional representations of the position and flight characteristics of all aircraft in the surrounding airspace. Working on the same principle as secondary surveillance radar, TCAS identifies and interrogates the transponder of any aircraft in the vicinity to determine whether its proximity (in terms of track, altitude, vertical speed or heading) poses a collision risk. The system automatically codes each threat and provides a series of visual and aural warnings to help pilots avoid collision.

Given the computer-mediated environment in which they work, the perceptual demands placed on pilots are considerable. They must continually synthesize accurate spatial awareness from a considerable amount of coded raw data, a task that requires training, skill, discipline and judgement in an uncertain and changing environment, together with quick, prudent decision-making based on a knowledge of the aircraft's systems and natural environment, crew capabilities and personal limitations. Pilots must remain 'ahead of the plane' in time and space to anticipate what they are likely to encounter in the short term and take actions to avoid potential problems. The maintenance of this spatial awareness requires continually monitoring the status, attributes and dynamics of the flight (including airspeeds, position, altitude, heading, ATC transmissions, TCAS returns and weather radar), while simultaneously comprehending their meaning and significance and projecting their status into the near future.

While computer code helps produce airspace on the flightdeck, the role of the human pilot remains crucial. As with ATC, experience and discretion are fundamentally important to the safe production of airspace, and pilots proactively negotiate the airspace through which they fly. For example, a pilot 'must not only comprehend that a weather cell – given its position, movement and intensity – is likely to create a hazardous situation within a certain period of time, but s/he must also determine what airspace will be available for route diversions, and ascertain where other potential conflicts may develop' (Endsley *et al.* 1998: 2). Thus, even if nominally following the same flightplan, no two aircraft use the sky in the same way, even though safety regulations dictate all manoeuvres fall within the boundaries of acceptable practice. Thus, as Dodge and Kitchin (2004) recognize, the production of airspace on the flightdeck is not universal or technologically determined, but contingent upon the embodied performances and practices of individual pilots, who use their discretion and experience to interact with nominally identical, yet subtly different, systems, equipment and environments.

So far, this chapter has explored some of the ways in which controllers and pilots mediate the production of airspace above the UK. Significantly, however, the ways in which airspace is used are often dictated, to a greater or lesser extent, by the topographical and socio-economic characteristics of the ground beneath it. The RAF choose to conduct much of their low-level flight training in mid Wales and the Scottish highlands, not only because of the challenging terrain but also because the noise associated with these operations will affect relatively few people. Similarly, at some airports, commercial aircraft may fly sub-optimal departure or arrival routes to avoid overflying densely populated urban areas. The following section explores how communities on the ground have started to challenge how the airspace above them is used.

Contesting the sky

Commercial air travel is becoming an increasingly emotive subject, and the debate surrounding who should benefit from, and, perhaps more importantly, who should suffer the impacts of aircraft noise and airport development has had a long pedigree. Given society's current socio-economic reliance on, and apparent 'addiction' to flying, this controversy appears to be intensifying as the relative cost of air travel declines and the number of flights increases. In the UK, passenger numbers have increased five-fold in the last thirty years, and forecasts suggest as many as 400–600 million passengers a year could be using UK airports by 2030 (Department for Transport 2003).

While air traffic controllers and pilots work to create airspaces that are safe for flight, they are also increasingly aware of their social and environmental responsibilities to reduce noise and emissions as far as possible. While the phenomenon of anti-airport protest is not new, current expansion plans and rising levels of public concern about aviation's contribution to climate change have caused the issue to rise up the political agenda. Some twenty-five anti-airport expansion groups are currently active in the UK, and range from small, local campaigns with limited membership to national pressure groups. Some of the larger organizations, including HACANClearskies (based at Heathrow) and SSE (Stop Stansted Expansion), have been instrumental in producing alternative understandings of airspace that challenge the dominant economic and operational discourses employed by airports, airlines and other pro-aviation lobbies (see Griggs and Howarth (2004) for a detailed study of the HACANClearskies campaign).

While the majority of campaigns oppose the development of new infra-structure, such as additional runways or new terminals, others are challenging changes to airspace that have resulted in commercial flights flying over their homes and communities for the first time. In 2005, the Dedham Vale Society won a High Court ruling against Stansted Airport and forced them to withdraw new flightpaths that, the group claimed, were ruining the rural tranquillity of 'Constable country' (Millward and Clover 2006). At East Midlands Airport (EMA), East Leicestershire Villages Against Airspace (ELVAA) also opposed plans to reorganize the airport's controlled airspace on grounds of noise, rural landscape despoliation and property devaluation.

In October 2003, EMA submitted an airspace change proposal to the CAA that sought to extend the area of controlled airspace around the airport and reorganize the way air traffic movements were handled. The plans involved amending existing approach and departure procedures, and re-siting the two holding areas or stacks to increase capacity and improve safety. While the plans were predicted to lessen the acoustic impact of aircraft operations on settlements in west Leicestershire and southern Derbyshire, a number of residents in east Leicestershire, who found themselves under the re-routed flightpaths, mobilized against the plans, believing they would cause unacceptable levels of noise pollution in a predominately rural part of the county (Staples 2004).

Following a public meeting in January 2004, a group of local residents formed ELVAA to raise awareness of the airspace change, stimulate public opposition, and act as a focal point of resistance. Significantly, ELVAA disputed the airport's claim that far fewer people would be subject to aircraft noise and claimed over 100,000 new people would be affected (Edwards and Farmer 2004). It also commissioned its own independent reports and noise surveys to challenge the claims put forward by the airport authorities. ELVAA quickly 'learned the language' of airport protest, and supporters lobbied local MPs, wrote letters of objection, and inundated the local media with their concerns. Anti-flightpath posters and messages of defiance also appeared on telegraph poles, hedgerows and village notice boards to try to raise awareness of (and galvanize support for) the campaign (Figure 6.5).

ELVAA's campaign was initially articulated in typical 'not-in-my-back-yard', or 'NIMBY', language, with spokespeople citing the loss of rural tranquillity and detrimental effects on quality of life that would result from the skies over east Leicestershire being turned into a '24 hour motorway for planes' (cited in the Oadby and Wigston Mail (2004)). In January 2005, with the airspace reorganization likely to go ahead, the group refocused their attention on trying to get the airport 'designated' under Section 78 of the 1982 Civil Aviation Act, which would place a cap on the number of night flights allowed at the airport, already the site of one of the UK's largest night-flying operations. Supporters of 'DEMAND' (Demand East Midlands Airport is Now Designated) as the group was subsequently renamed, argued night-time freight flights disturbed their sleep and breached their 'right' to peace and quiet.

In May 2005, the airspace change was implemented, and the new flightpaths undoubtedly changed the acoustic environment of east Leicestershire. The emotional upset this caused for some individuals helped create a territorial identity for ELVAA and DEMAND, where acceptance into the group was

Figure 6.5 Manifestations of protest: ELVAA's roadside poster campaign in east Leicestershire sought to energize local resistance to the airspace change.

determined by the ability to hear aircraft noise and a willingness to protest against the perceived injustices of authority. While the majority of objections and complaints ostensibly employed the familiar rhetoric of rural landscape despoliation and feelings of being 'overwhelmed' by noise, others implied they did not understand why the airspace reconfiguration was considered necessary. Some suggested aircraft could be routed over East Anglia but, owing to the number of military air traffic zones and other areas of intense aerial activity over the region and the fact the majority of aircraft fly on a north-south trajectory, this suggestion was impractical. Though EMA remains, at present, undesignated, ELVAA supporters used their lived experience of the airspace change to produce alternative notions of airspace that are largely incompatible with those of the airport.

Conclusion

UK airspace is a product of numerous interlocking geopolitical, economic, environmental, social, technical and commercial practices that operate at a variety of spatial scales and manifest themselves in different ways in different places through time. As a consequence, the existing airspace structure is a compromise, designed to ensure all users, from the Royal Air Force, to commercial airlines, air ambulances and private pilots, can access it, albeit in ways that are often restricted. While access to, and use of, sovereign airspace is controlled by the state, individual citizens are relatively powerless to dictate how the airspace above their personal property is used. Evidence of anti-airspace protests at EMA, Stansted and elsewhere in the UK suggests that many individuals are becoming increasingly intolerant of aircraft noise and disturbance.

This chapter has illustrated that airspace is simultaneously produced 'from above' by controllers and pilots and challenged by people 'on the ground' who oppose its use. It has suggested that airspace must be conceptualized not as a 'tunnel' of mobility in the sky, but as an important social space in its own right, mediated by numerous different users and agencies and imbued with meanings, values and significance that we are only beginning to understand. As the controversy surrounding the growth of air travel intensifies, debates about the 'acceptable' use of airspace will become more common as new flightpaths are introduced to handle growing numbers of aircraft. While the technology exists to make the existing airspace structure more efficient by eliminating circuitous routes and enabling aircraft to fly a direct line from A to B, it is questionable whether the political will, and the finance, to enact such changes exist. Until they do, we must devise new ways of safely accommodating growing volumes of air traffic in an airspace system designed in the previous century. As one controller pertinently remarked, 'controlling planes is easy – pilots generally do what they're told. It's making space in the sky for them that's difficult.'[5]

Acknowledgements

I would like to thank Martin Weir at bmi for his assistance and Iain Millan and Brian King at Navtech/European Aeronautical Group for permission to reproduce the airspace charts.

Notes

1   Source: Personal communication, Air Traffic Controller, Swanwick. Anonymous by request.
2   Source: Personal communication, Airspace Planner, Swanwick. Anonymous by request.
3   See Duke (2005) for a detailed description of the different classes of airspace.
4   Source: Personal communication, Air Traffic Controller, EMA. Anonymous by request.
5   Source: Personal communication, Air Traffic Controller, EMA. Anonymous by request.

Bibliography

Balfour, J. (1994) 'The changing role of regulation in European air transport liberalization', *Journal of Air Transport Management*, 1(1): 27–36.
Beaty, D. (1991) *The Naked Pilot: The Human Factor in Aircraft Accidents*, London: Methuen.
Brittin, B. H. and Watson, L. B. (1972) *International Law for Seagoing Officers*, 3rd edn, Annapolis Maryland: Naval Institute Press.
Burney, C. D. (1929) *The World, the Air, and the Future*, London: George Allen and Unwin Ltd.
CAA (2007) *Civil Aviation Authority airspace infringement statistics*, Online. Available at www.flyontrack.co.uk (accessed 20 July 2007).
Calder, S. (2002) *No Frills: The Truth Behind the Low-Cost Revolution in the Skies*, London: Virgin Books.
Castells, M. (1996) *The Rise of the Network Society: The Information Age: Economy, Society, and Culture*, vol. 1, Oxford: Blackwell.
Dargon, J. (1919) *The Future of Aviation*, trans. P. Nutt, London: David Nutt.
Department for Transport (2003) *The Future of Air Transport*, White Paper Crn6046, December 2003, London: HMSO.
Dodge, M. and Kitchin, R. (2004) 'Flying through code/space: the real virtuality of air travel', *Environment and Planning A*, 36: 195–211.
—— and —— (2005) Code and the Transduction of Space, *Annals of the Association of American Geographers*, 95(1): 162–80.
Duke, G. (2005) *Air Traffic Control*, 9th edn, Hersham: Ian Allan.
Edwards, K. and Farmer, B. (2004) '100,000 people live beneath flight paths', *Leicester Mercury*, 31 July 2004: 6–7.
Endsley, M. R., Farley, T. C., Jones, W. M., Midkiff, A. H., and Hansman, R. J. (1998) *Situation Awareness Information Requirements for Commercial Airline Pilots*, International Centre for Air Transportation, Cambridge, Massachusetts: MIT Press.
Faith, N. (1996) *Black Box: Why Air Safety is No Accident*, London: Boxtree Macmillan.

Finch, R. (1938) *The World's Airways*, London: University of London Press.

Fortson, D. (2007) 'Going like a dream', *The Independent on Sunday Business*, 22 April 2007: 6–7.

Graham, S. (2005) Software-Sorted Geographies, *Progress in Human Geography*, 29(5): 562–80.

—— and Marvin, S. (2001) *Splintering urbanism: Networked infrastructures, technological mobilities and the urban condition*, London: Routledge.

Griggs, S. and Howarth, D. (2004) 'A transformative political campaign? The new rhetoric of protest against airport expansion in the UK', *Journal of Political Ideologies*, 9(2): 181–201.

Heath, C., Hindmarsh, J. and Luff, P. (1999) 'Interaction in Isolation: The Dislocated World of the London Underground train driver', *Sociology*, 33(3): 555–75.

Jones, P. (2005) 'Send Three and Fourpence', *focus on commercial aviation safety*, United Kingdom Flight Safety Committee (UKFSC), Autumn 61: 5–7.

Lawton, T. (2002) *Cleared for Take-off: Structure and Strategy in the Low Fare Airline Business*, Aldershot: Ashgate.

Millichap, R. J. (2000) 'Airline Markets and Regulation', in P. Jarrett (ed.) *Modern Air Transport: Worldwide Air Transport from 1945 to the Present*, London: Putnam Aeronautical Books: 35–52.

Millward, D. and Clover, C. (2006) 'Constable country sees off intruders into its airspace', *The Daily Telegraph*, 2 January 2006, Online. Available at www.telegraph.co.uk/news (accessed 13 January 2006).

National Air Traffic Services Ltd (2005) *The Start of Commercial Aviation, 1919–1929*, Online. Available at www.nats.co.uk/library/history2.html (accessed 10 October 2005).

Norris, G. and Wagner, M. (1996) *Boeing 777*, Osceola, Wisconsin: Motorbooks International.

Oadby and Wigston Mail (2004) 'Flight Fight', 17 June 2004: 4.

Petzinger, T. (1995) *Hard Landing: How the Epic Contest for Power and Profits Plunged the Airlines into Chaos*, London: Aurum.

Staples, S. (2004) A rude awakening for peaceful corner of the countryside, *Leicester Mercury*, 4 August 2004.

Thrift, N. and French, S. (2002) 'The automatic production of space', *Transactions of the Institute of British Geographers*, NS 27(3): 309–35.

Veale, S. E. (1945) *To-morrow's Airliners, Airways and Airports*, London: Pilot Press.

7   Around the world in 80 airports

Photo essay by
Ross Rudesch Harley

Imagine you got on a plane, and didn't stop – just flew from one city to the next, without ever really touching the ground. Imagine a journey around the world, where you never leave the realm of the airport.

This project can be read in part as a response to the conceptual art practice of the 1970s that dealt with nature and culture through the creation of work about the relation between time, space, distance, geography, measurement and experience. A modification of Richard Long's well-known nature walks, this project is a formal and media-filtered photo-essay that describes the non-space and experience of global airports and their most elemental materials and flows.

The images presented here have been gathered over a period of five years and are at once documentations, deconstructions and interventions in the environment of airports. Referring back to the nineteenth-century dream of circumnavigating the globe in just 80 hours as popularised by Jules Verne's *Around the World in 80 Days*, this project exchanges time for space. The network of airports visited creates a personal map of the world that turns out to be shared by many millions of other travelers who find themselves tracing similar routes in the course of their daily life.

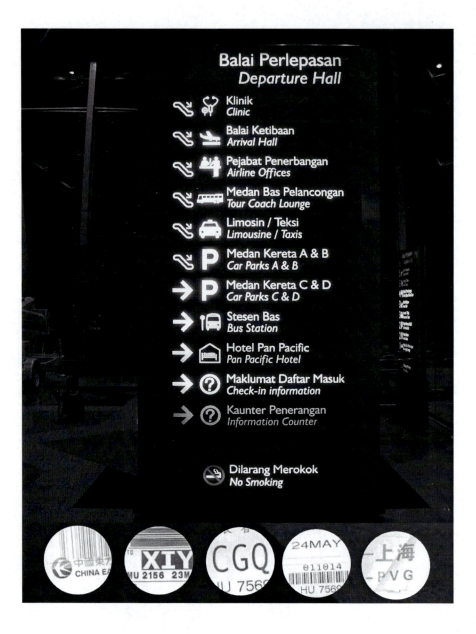

The global air terminal network is a system of interconnecting tunnels and tubes, wayfaring sytems, semantic webs and motion architecture. We enter and leave this travel system without thinking twice, such is the technical finesse and ubiquity of this network. It moves with us from our doorstep to the terminal, engulfs us, surrounds us with images-as-things, words-as-things and signs-as-nav-systems. It is a viral architecture of exponential reproduction, a briefly habit-able image-bubble that trades under three-letter codes and 'universal' barcodes and infographics.

In the nineteenth century, the very thought of traveling 'around the world in 80 days' generated considerable excitement and debate about the liberatory prospects of modern technology and science. By the mid-twentieth century, Jules Verne's classic had been translated into paperback, comics, popular music and wide-screen feature films. Would the world really change if time and distance could be so compressed by media and travel machines? The answer is a resounding 'yes'.

The advent of the jet age in the 1950s saw *Time-Life* magazine revise Verne's no-longer-fanciful proposition of circumscribing the globe in a handful of weeks. Entitled *Around the world in 80 hours*, the 1956 summer edition of the magazine devoted an entire issue to the marvels of the modern age of flight. American illustrator Edward Weep accompanied US Congresswoman Claire Booth Luce on an 80-hour whirlwind world tour (courtesy of *Time-Life*, Pan Am and Intercontinental Hotels).

The late 1950s subsequently marks the moment that air travel moves from the preserve of a venturesome elite to the domain of the masses. Nearly fifty years after *Time-Life*, I decided to embark on my own Vernian adventure, aiming to travel not so much against time as through space.

My idea was to buy a round-the-world-ticket and to photograph as many airports as I could on that ticket. The plan was to go *Around the World in 80 Airports*, sleeping on the plane and photographing terminals between flights. It seemed simple and straight-forward to me, but everyone, including my travel agent, couldn't believe it. Was I really going to spend a month in the air without ever reaching a destination? What would my body say? And wouldn't it look pretty suspicious to all those post-9/11 customs or security officials? As it turns out, I had set myself an impossible task. Round-the-world tickets allow a maximum of sixteen destinations, and so my project ended up being split over a number of different trips over a five-year period. This photo-essay is based on material gathered from those trips.

The air corridors that lead from one airport to the next create a very particular aerial and terrestrial image of the globe, which in this instance is made entirely from itineraries limited to airports I have personally visited. These flightpaths have been 'sketched up' in Google Earth, and then rendered in 3D modeling software to create transnational 'airbridges' that slice across time and space. Together, these tubular-looking flightpaths create a dynamic spatio-temporal image of our world girded by spindly tendrils that terminate at the runway. The project also presents a 4D perspective, whereby different journeys taken over time are represented in itineraries that exist in the one time/space of the visualisation.

With its abstracted 3D extrusions of terminals, runways and flightpaths, the whole system looks like a techno-rhizome, with its complex system of flightpath-aerials and terminal-runway-roots. This project has also been conceived to be presented as multiscreen video projection/loops and interactive components (at the moment constructed in Google Earth). The extensive photo documentation from each of the airports is stored in large photo-sets at Flickr, and can be displayed in relation to the main 3D work (on the web, as photographic prints and in a variety of publications).

All photos Ross Rudesch Harley. Google Earth hacks and 3D renders by Mister Snow and the House of Laudenam, 2006/7. For more information visit stereopresence.net.

Part III

The social life of air travel

8 Airborne on time

Peter Peters

Introduction

There are multiple ways of knowing an airport. One can study its architecture as an expression of the ideals of modernity and 'flow' (Pascoe 2001; Misa *et al.* 2003) or reflect on the relativity of its (im)mobilities (Adey 2007). The airport has been portrayed as the quintessential 'non-place' (Augé 1995; Castells 1996) and as the locus of the merging of global and local politics of mobility and sustainability (Kesselring 2006). Yet, in all these accounts, relatively little attention has been paid to the actual work that is required to render reliable worldwide air travel on a daily basis. In this chapter, I focus on the practicalities of everyday air travel from the perspective of a large airline, the French/Dutch company Air France-KLM.[1] This study is based on ethnographic research at Amsterdam Airport Schiphol (AAS) that started a number of years ago from a question that is at once simple and highly complex in its repercussions: where is time in an airport?

To see how airport time can open a perspective on the daily practice of air travel, we cannot reduce it to clock time. Instead, we have to examine how dispersed events and processes are synchronized. All these different times come together in the moment an aircraft becomes airborne. To arrive in Boston seven hours after leaving Amsterdam, a heterogeneous order must be created. Aircraft, passengers, crew, baggage, fuel and catering have to be at the right place at the right time, and this can only be achieved by carefully synchronizing a complex chain of actions. This synchronization is achieved in part before take-off. For example, an airline's flight plans order both the maintenance schedules of the aircraft in its fleet and the bookings of its passengers made by travel agencies and through websites. Whether the planned order will become the real order of events, however, can only be known when an aircraft becomes airborne. Passengers may arrive late at the gate, congestion in the airways may make a different and longer route necessary, or loading the baggage onto an aircraft may run behind schedule. Disruptions in the planned order of an airline's daily operations can have a variety of causes and usually end in delays. If an airline such as KLM wants to be punctual, the planned order of its daily flight schedule has to be repaired continuously by countering contingencies that may cause disruptions (La Porte 1988). How can KLM prevent gridlock from arising in its system?

This chapter examines how this is achieved. It is based on field work in several locations at AAS: KLM's departure hall, the hub control centre (HCC), where all hub operations are managed, and KLM operations control centre, where the daily operations of its global network of connections are monitored. My fieldwork included observations in each of these locations and interviews with KLM employees.[2] After describing and analyzing their daily activities, it will then become clear that repairing the heterogeneous order that creates a flight is a practical achievement that assumes the ability to coin 'exchange', as one of the KLM employees expressed it. Exchange is what keeps the operation of any given flight going. It comes in several currencies and mediates between the rigid realities of plans, clocks and protocols and a continuous flow of contingencies and everyday problems (Peters 2006).

On the platform

On a Thursday morning in June 2007, I stood in the cargo hold of a KLM Boeing 747 Combi. The aircraft was scheduled for departure to Mexico City in about two hours. A number of cargo pallets had to be loaded, one of which was larger than usual. The pallet was lifted to the level of the cargo deck and rolled slowly into the hold. Personnel from Ground Services tried to position the pallet, using the metal locks on the floor. The locks were situated at regular intervals, but, in order to fasten the oversized pallet, some of the locks had to be readjusted. The engineer began hammering the lock and eventually managed to loosen it enough to adjust it to the dimensions of the pallet. It looked like a routine act. The standardized dimensions inside the cargo hold, consisting of neatly aligned rows of pallet locks, are momentarily reordered to accommodate the non-standard size of the pallet. The experienced KLM platform personnel know that cargo locks can get stuck and therefore always take a hammer in case they need to adjust them. Without it, the loading process could run behind schedule, causing a delay that in turn might ripple through the day's flight plans of multiple aircrafts and cause disruptions in the network's operations.

Standing on the cargo lift, I was lowered down to the platform. There I met the Team Leader Turnaround, who is responsible for loading the aircraft and supervising the boarding of the passengers, with the Gate Agent. We walked to the foot of the boarding bridge, climbed the stairs, passed the security officer inside the bridge and entered the plane. In the cockpit, the Team Leader handed over an aircraft data load sheet to the first pilot, one of the standard procedures in the aircraft turnaround process. Through the cockpit windows we could see passengers walking through the bridge to board the aircraft. We left the cockpit and walked in the opposite direction to the gate. The Team Leader and the Gate Agent exchanged some details on the boarding process. Back on the platform, we were informed that all doors were closed, and the aircraft was ready to be pushed back. The boarding bridge slowly moved backwards. Suddenly the Team Leader's walkie-talkie began to crackle. It was the Gate

Figure 8.1 The Team Leader Turnaround on the platform being apprised of a missing passenger shortly before push-off of a flight to Mexico City (photo taken by the author with permission from the Team Leader Turnaround).

Agent explaining that there was a passenger missing who was now standing at the gate. The Team Leader was not amused. It meant a potential delay of at least ten minutes. The boarding bridge was reconnected to the plane, and the late passenger rushed through the bridge into the aircraft.

Passengers who check in but do not show up on time at the gate are called in airport jargon '-LMCs' (minus last minute changes).[3] -LMCs create trouble because, for security reasons, passengers and their baggage must always be on the same plane. Unaccompanied baggage must be taken off an aircraft, which can cause significant delays. The sooner KLM employees know a passenger is missing, the quicker they can react. For this reason, KLM moved up the time a passenger for a European flight must report to the gate to within ten minutes of boarding time. The additional time provides more space to act while remaining within the flight schedule. However, the building in of such buffer time required the extension of the duration of the total air journey. With the delay of the flight to Mexico, the Team Leader had to assign a code to the event that caused the delay. Delays caused by passengers who do not show up at the gate are represented by code 03. Code 03 is one of ninety-nine delay codes that are printed on a card which all KLM employees have close at hand.[4]

The work of the KLM employees on the platform and at the gate is characterized by a tension. They have to comply with protocols, rules and conditions, but they also need the 'rule space' to solve a myriad of problems, while at the same time smoothing out the disruptions they can create in the daily operation of flights as a whole. The platform is just one of the many places that make the heterogeneous order of a flight. Other places are the departure hall, maintenance hangars, the baggage handling spaces, the air traffic control tower, the kitchens of airport catering. As airports are linked in a network of global connections, these 'duty areas' are spatially distributed, not only in the airport itself, but all over the world. The intriguing question, then, is how is it possible to coordinate the actions of countless individuals who are in so many places? How is it possible to create an overview of these dispersed personnel? How can an airline make sure that adjustments in the heterogeneous order of any given aeroplane's passage in one place do not lead to disruptions in another? To put these questions in perspective, I move the site of my field work to the HCC of KLM.

In the hub control centre

If a Team Leader Turnaround encounters a problem in the aircraft turnaround process that may potentially disrupt the on-time departure of an aeroplane, he has to report this to the Duty Area Manager in the KLM HCC. The HCC is part of the division of Ground Services that is responsible for all ground handling of passengers, baggage and freight for KLM at AAS. The airport has five 'duty areas', two for European flights, two for intercontinental flights and one for commuter flights (Cityhopper). The Duty Area Manager supervises the work of a number of turnaround processes and communicates with the Team Leaders and the Gate Agents. The supervision of all activities on the hub is the responsibility of the Duty Hub Manager.

The HCC is housed in offices that overlook Departure Hall 2 at Schiphol, which is designated primarily for KLM's use. On a Saturday morning, the Duty Area Manager of intercontinental flights prepares for a 7.30 a.m. conference call. He briefly recalls the weather forecast – a few thunderstorms in the afternoon – and continues with the main facts and figures of the day, such as the number of departing and arriving passengers. Two flights are designated as 'star flights', meaning that it is crucial that they depart on time. Another problem for this day is the large number of time slots all over Europe due to weather conditions that have decreased airway capacity. One by one, representatives of all the departments involved in Ground Services report on their status and comment on expected or potential problems. These may be understaffed towing services, planes that have been damaged and are still in the hangar, or problems in gate planning.

The Duty Hub Manager explains that his work is all about anticipating potential problems and disruptions:

My planning horizon is six to eight hours. If I know there are understaffed departments, I need to know the consequences. What flight numbers are involved? If we need to make six towing movements, but Towing Services can only do four, we have to decide which flights we have to delay. In order to coordinate and steer the operations, we need the information from the shop floor and we need it as early as possible. That is why we have put a lot of effort in generating timely status reports by the people on the platform, in the departure halls, and at the gates.

(Interview with the Duty Hub Manager)

A continuous flow of information enters the HCC. The information is presented on computer screens, through 'old technologies', such as telex and walkie-talkies, and through telephones. The soundscape in the control centre is dominated by people talking and laughing, phones ringing and matrix printers rattling. Flight numbers are constantly exchanged as small capsules of information:

We've got the slot for the 427. I will try to put it at 14:35. At Fox 9 we still have the 641 waiting. Was supposed to be ready. Flow Control issued a new slot for the 887. Means it's heavily delayed. But is his arrival time actual? At Fox 5, the 587 is ready for push back, but I cannot see it. We have some late pallets for the 476.

(Recording of Duty Area Manager)

The people in the HCC use several mainframe information systems that together provide detailed real-time information on the status of flights. By typing in short commands, KLM personnel all over the airport can request information on any flight.[5]

If problems arise, how much room for manoeuvre is there? Who decides if a departing flight can wait for the passengers from a connecting flight that is delayed? As it turns out, the control centre at the hub overlooks and coordinates a wide range of processes, all of which have to connect and integrate to render an on-time performance of the flight schedule. The rule space of the HCC can be summarized as 'V + 15', or fifteen minutes past the scheduled departure time for intercontinental flights, and 'V + 5' for European flights. If a disruption causes a flight to be delayed for a longer time, the hub has to contact the Operations Control Centre, which supervises KLM operations worldwide.

Creating an overview

The building in which KLM's Operations Control Centre resides opened in September 1999 and is located at Schiphol Airport-East, where the original airport was constructed in the 1920s. It is now a business district that houses airline-related companies. The Operations Control Centre has, next to its Back

Office where flight schedules are assessed, a Front Office where processes at the network level are monitored in real time on the day of operations. The Front Office of the control centre is a spacious, climate-controlled hall. At the front end of the Front Office is a massive video screen that spans the width of the hall. It was intended to show all the present flights in real time, but is currently not in use because of cost-cutting measures. Below the video wall is a row of digital clocks that display, in green, the current time in different cities around the world. In the middle is a clock showing in red the Universal Time Coordinated (UTC) (see Figure 8.2).[6] The massive hall contains several rows of desks, each of which has two or three computer screens, other communications equipment and paper. Not all the desks are occupied, however. People are walking, talking and using their telephones. At the front of the hall there are two televisions screens relaying a live CNN broadcast and Teletext.

The Front Office is on duty 24 hours a day. People work in three shifts. Every morning the new day of operations starts with what is known as the 'morning prayer' at 8 a.m., a short meeting to evaluate the previous day and prepare for the upcoming one.

All the KLM divisions that are relevant to the daily operations of the network are represented in the Front Office. The most important employees there are the Duty Manager Operations and three Operations Controllers (OCs), who

Figure 8.2 The front office in the KLM Operations Control Centre (photo taken by the author with permission of those photographed).

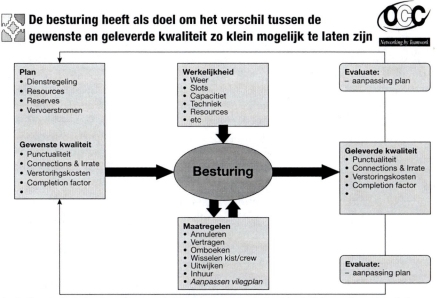

De besturing heeft als doel om het verschil tussen de gewenste en geleverde kwaliteit zo klein mogelijk te laten zijn

A = de wijz waarop de dienstregeling volgens oorspronkelijk plan moet worden uitgevoerd en de punctualiteit, connecties, verstoringskosten, irrate en aantal vluchten die hetplan op zouden moeten leveren
B = de operationele omstandigheden die, het plan (kunnen) verstoren
C = het kiezen ult een van de maatregelen om de gevolgen van een verstoring te beperken
D = de gekozen maatregel
E = de waargemaakte punctualitiet, (no-)connecties, bagage Irrate, verstoringskosten, aantal vluchten, etc

Figure 8.3 A chart of the flow of operations in KLM's Operations Control Centre (from an internal KLM Operations Control Centre leaflet, reproduced with permission from KLM).

always include one Senior OC. Although the Duty Manager Operations is at the top of a hierarchical line of staff, he is not involved in daily decisions. These are made by the OCs, who are responsible for the daily performance of the local, European and intercontinental flight schedules of the network's operations. The OCs are supported by a representative from every division that is relevant to the daily operations of the network. Before moving to the new Operations Control Centre – which cost €55 million – in 1999, the OCs were housed in separate buildings on the airport complex premises. But KLM believed that face-to-face interactions among employees would improve its ability to respond to contingencies and disruptions in its airline network. It remains confident that the costly investment will eventually pay off.

A control protocol contains instructions on how the Operations Control Centre 'controls the operation'. The protocol describes lines of communication, decision-making procedures, rules and corporate culture. It states that the central goal of the operation is to 'minimize the difference between projected and realized quality in the performance of the flight schedule'. Control through the Operations Control Centre covers the entire process, from the acceptance of the plan and the stated level of quality, to responses to surrounding factors, proactive measures and their communication to relevant members of the organization.

Figure 8.3 is taken from a KLM leaflet that is on every desk in the Front Office. It illustrates how the Operations Control Centre defines its own steering strategy (*Besturing*). Its point of departure is the organizational plan, which includes not only flight schedules, but also the availability of planes, planes in reserve, crew and the expected number of passengers. A desired level of quality is prescribed, which incorporates an acceptable deviation from the flight schedule. However, no absolute indication of the desired level of quality can be given, because it depends on external circumstances, such as weather conditions, capacity and market strategies. In defining the desired and realized levels of quality, departure punctuality, delay costs, irregular baggage handling, cancellations and missed connecting flights are taken as the key performance indicators. There are all kinds of factors that enter into the realization of the plan which the Operations Control Centre manual describes as the ability to 'put a spoke in the wheel': weather, flight slots, technical failures, shortage of capacity on runways and shortage of personnel. If the desired level of quality is threatened, then 'proactive measures' must be taken to prevent further disruptions. A plane or crew must be changed, an aircraft flown to an alternative airport, extra capacity rented, a plane's internal configuration changed or a flight delayed or cancelled.[7]

The Front Office makes an overview of activities that are both spatially and temporally distributed possible. Suchman (1993) has pointed out that control centres, as 'centres of coordination', are designed to maintain a contradictory state of affairs. In order to function, a control centre has to occupy a stable site to which physically dispersed agents can connect, but it must also be able to access all other relevant situations that are distant in space and time (1993: 115) (see also Suchman 1997). The type of control that is enacted in KLM's Front Office is a practical achievement rendered by both creating an overview of physically distributed situations and changing them in relation to each other. This achievement, in turn, rests on (a) real time coordination of cooperative work that is both spatially and temporally distributed; and (b) a redistribution of accountabilities and responsibilities in such a way that the power to act in distributed situations is limited to a small number of employees in one place.[8]

Breakdown in Barcelona

On the computer screen that sits in front of the Operations Controller, all KLM's flights for the day are displayed as coloured horizontal bars. The horizontal axis represents time, with each division, indicated by a yellow dotted line, designating one hour in a 24-hour day time frame. A vertical blue dashed line marks the passage of UTC time (see Figure 8.4).

In the far left-hand column, the planes in KLM's fleet are listed. The yellow and green horizontal bars represent the flights of each aircraft in the fleet on

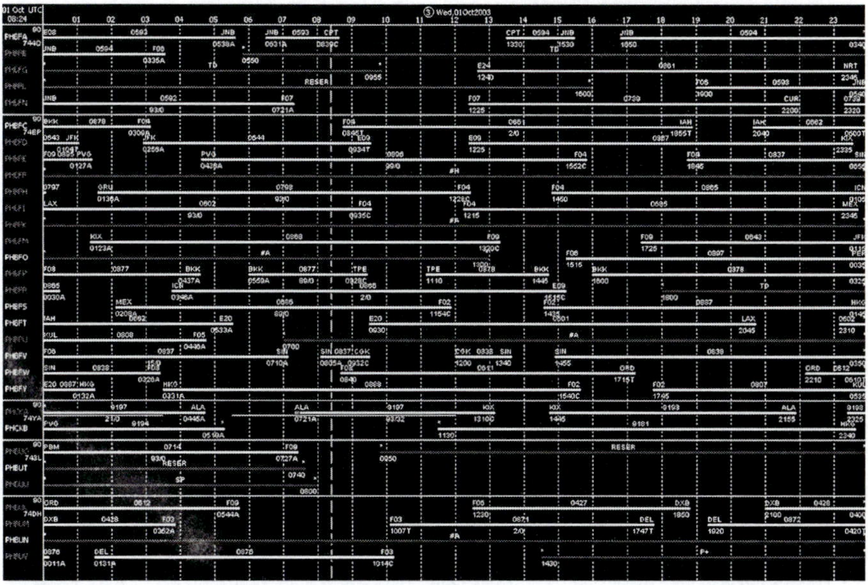

Figure 8.4 A typical screen image of Flash, a computer program that shows the KLM flight schedule in real time.

a given day, in this case Wednesday 1 October 2003. Red bars mean that an aircraft is scheduled for maintenance. The bars can be shorter or longer, depending on the amount of time an aircraft is 'off blocks', the time between its departure from the gate of origin and its arrival at the gate of its destination. Departing flights, indicated by the yellow bars on the screen, have odd numbers; arriving planes, shown in green, even numbers. The block times for each flight are given under the bars. If a flight is delayed, part of the bar will become red. Above the bars are the departure or arrival gates, and, for departing flights, the destinations designated by their international abbreviations. By clicking on a bar on the computer screen, the OC can get all the information he needs for any flight: the gate number at Schiphol Airport, the registration code, the names of the crew members, the plane's recent maintenance history, etc. He can move a bar to the left to indicate a delay, or to the right for flights travelling ahead of schedule. He can also move the bar upwards or downwards to indicate a change of plane. With simple mouse clicks, an OC can adjust the flight schedule, but he knows his decisions may have a great impact in many places in the airport:

> A change of gate has to be communicated to the passengers, their baggage has to be moved to another location, catering, fuelling, everything has to be adapted and re-synchronized. They are often not very happy, but there

is nothing I can do about it. Here we have the overview, [out there] they only see their own processes.

(Interview with OC Europe)

When a telephone rings, the OC Europe turns the speaker phones on so everyone can hear the conversation. The pilot of a 737 at Barcelona airport reports that there is a hole in the body of his plane. A luggage tractor has taken a turn too sharply and damaged the aircraft's aluminium skin. The OC organizes a telephone conference to discuss the problem. By pushing on a display in front of him, he is able to talk simultaneously to the pilot, the engineers and the Service Manager in Barcelona and the KLM Maintenance Division who are sitting behind him in the Front Office.

The damage to the Barcelona plane is serious enough to delay its return to Amsterdam indefinitely. This creates a number of problems that the OC Europe must resolve. Even if the plane can be repaired, safety regulations require that damaged planes must be taken out of service. This means that it must be flown back to Amsterdam empty. All passengers will have to be re-routed on other flights to Amsterdam, and another aircraft must also be found for those who were booked through to Hamburg, the flight's final destination. The flight and cabin crew must be changed, because it is probable that the delay will mean that they will exceed the number of hours they are legally permitted to work. Since they were also scheduled for a subsequent flight from Amsterdam, another crew will have to be found to replace them on that flight as well. The OC looks at the computer screen for a plane that will be used for the flight to Hamburg. A 737 is available, but is scheduled for maintenance. He must find out if its maintenance check can be postponed. He contacts the Duty Maintenance Manager who confirms that it can be. But he must also check if this 737 has enough seats and freight capacity for the Hamburg flight.

The OC Europe moves the yellow bars on his screen. The problems that the damaged Barcelona plane caused have been solved. The adjustments are put on the telex and sent to all relevant divisions. All electronic systems, from the computers in travel agencies to the public displays at Schiphol Airport, immediately register the new reality that the OC Europe has negotiated.

Work in real time

The breakdown in Barcelona illustrates how the overview of KLM's operations created in the Front Office enables disruptions that occur somewhere in the planned order of an aircraft's schedule to be controlled. Real-time operations require that problems be solved when they occur. KLM cannot stop its operations to design a solution.[9] How, then, are real-time decisions made? We can answer this question by examining the control protocol in Figure 8.4, in which there is a clear distinction between an a priori state of operations (the plan, the desired level of quality), a reality that can be influenced by taking measures, and a post hoc evaluation, when the quality of the decisions is measured. Although the chart reflects the way the Front Office staff conceptualize their work, it does not make clear how KLM employees actually repair the planned order of the passage in the moment a contingency occurs. What concrete steps do employees have at their disposal? Who decides when to deviate from the protocol?

The Front Office protocol represents possible situations in which action is required (bad weather, a lack of slot times, a lack of runway capacity, technical failures, etc.). But it cannot predict disruptive events before they occur. According to Suchman (1987), plans do not anticipate action, but rather plans and actions are interwoven. Every action is situated in the sense that, no matter how thoroughly planned it is, it will be subject to contingencies. The circumstances of our actions can never be fully anticipated and, thus, undergo continual change. Suchman does not deny the common-sense idea that actions can be planned, but claims that they can never be planned in the strong sense that cognitive science would have it.

> Stated in advance, plans are necessarily vague, insofar as they must accommodate the unforeseeable contingencies of particular situations. Reconstructed in retrospect, plans systematically filter out the particularity of detail that characterizes situated actions, in favour of those aspects of the actions that can be seen to accord with the plan.
>
> (Suchman 1987: ix)

In the Front Office this means that the relationship between any given protocol and how things play out on the ground cannot be predetermined. Instead, considerations and judgements must continually be made in view of changing circumstances. The Front Office staff rely on protocols and procedures, but because every disruption introduces a complex of contingencies, they will always need 'rule space' to be able to react. The results of their decisions will not be an exact performance of a previously formulated protocol, but will always deviate more or less from it. The flow charts of the control protocol are thus part of ongoing and emerging situations, rather than programmes preceding them.[10]

This also becomes clear in the use of a software program called the Traffic Flow Management Decision Support Tool. It can predict the consequences of any given decision by computing the real or estimated costs of delaying an aircraft from booking information. One of the OCs said he used the tool to see what the possible consequences of a decision might be, but more than once knew from experience that the outcome would actually be different. The Support Tool offers a means of predicting the cost of any given decision. But since the tool works with fixed inputs, an OC's experience is also necessary.

> The program may indicate that we need 50 minutes to turn around the plane, that is the time it takes to prepare the plane for taking off again

after we have landed in one of our destination airports. But I know the
people who work there, and know that they can do it in 30 minutes.

(Interview with an OC)

As we have seen from following the platform personnel in the HCC and the
Front Office around, problems have to be solved as they emerge, and operations cannot be stopped in order to design solutions. Any measure taken to
remedy a disturbance will have an immediate effect on all other connected
processes. Situated action can only partially account for this kind of complexity. The concept of 'improvisation' is useful here because it adds a more
explicit temporality to the situatedness of actions (Crossan 1998). As Ciborra
has pointed out, the moment of improvisation suggests the immediacy, idio-
syncrasy, as well as situated character of acquiring access to and deploying
resources in any given situation (Ciborra 1999: 84; Ciborra 2000). However,
the term improvisation also captures the creative character of acting in these
situations. The new emerges as elements within an existing order that are given
a different function or meaning in the course of time. Ciborra (1999: 84) refers
to this process as an 'ecology of improvisation' in which, for example, the
instruments and artefacts in a situation are 'annotated', if not 're-invented'.
Translated to KLM's Front Office, this means that an employee must be able
to oversee an emerging situation *and* to act in this situation on order to
solve a problem.

Concepts such as situated action and improvisation make it possible to
explain how Front Office staff solve emerging problems and repair the heterogeneous order of a flight schedule in real time. They account for the 'coupling'
of a plan and an action, while emphasizing the temporal and creative character
of these processes. The concept of 'situated action' accounts for the contingent
character of situations and the need constantly to re-evaluate a situation on the
basis of plans that are themselves undergoing change in the action. But it
restricts action to the cognitive skills of pattern recognition, experience and tacit
knowledge. And it is not clear how new actions are possible. In contrast,
improvisation takes the emergence of new actions as a product of creativity.
The new emerges when new meanings, uses or functions are given to elements
in the existing order. However, improvisation assumes an actor's autonomy:
to improvise one has to be *allowed* to restructure the order. The question of
who is allowed to improvise and who is not remains unanswered. Conceptualizing airport time as the result of repairing the heterogeneous order of a flight
schedule makes it possible to ask *how*, in the departure hall and the Operations
Control Centre, spatio-temporal orders are produced and related to one another,
and what consequences this has for the distribution of the power of decision.

Coining exchange

Sitting behind his computer display, the OC for Europe can delay flights by
shifting the green horizontal bars to the right or change planes by moving the

red vertical bars up or down. The screen image in Figure 8.4 represents the complexity of daily flight operations. It not only provides an overview, but also the possibility to act in geographically dispersed situations.The OC Europe solved the loss of the damaged Barcelona plane by using a 737 that was scheduled for maintenance. But he could only achieve this by consulting the Maintenance Division to determine if the plane could be taken off its maintenance schedule.

> This requires great precision. You cannot just use a plane that is scheduled for maintenance. Postponing maintenance may lead to problems in a few days, because the maintenance has to be done anyway. But in this case, I could reschedule the maintenance schedule in such a way that we will not have a problem later in the week. This plane was my exchange for today. I have now used it. If another plane is damaged, my problem will be harder to solve.
>
> (Interview with OC Europe)

I take the OC Europe's use of the term 'exchange' as a point of departure for understanding the practical achievement of repairing a flight schedule in real time. By postponing the 737's scheduled maintenance, the OC Europe made sure that the plane could be incorporated into the planned flight schedule when the problem in Barcelona occurred. He repaired the temporal order of flight KL1668 by changing the temporal order elsewhere in the operations network, in this case, in the Maintenance Division. If such a contingency had occurred one day earlier or later, it might not have been as easy to make this plane available.

In overseeing KLM's fleet, the parameters of the actions the Front Office staff can take are defined by a combination of protocols, codes and routines. KLM staff operate within these parameters, but they also are constantly renegotiating them. Solving problems such as a broken down plane can only be achieved by having a certain amount of exchange at one's disposal. The concept of exchange embodies a form of agency that enables OCs to link entities from distributed networks of time at any given moment, thus recreating the heterogeneous spatio-temporal order of a flight schedule *in real time*. To put it another way, in repairing the spatio-temporal order of a flight schedule, it is not the amount of clock time that furnishes the agency to do so, but the character of the exchange at hand.

KLM staff on the platform, in the departure hall, in the Hub Control Centre and in the Front Office are confronted with a number of situations in which they must have exchange available or be able to coin it on site in the moment a disruption arises. Exchange comes in different currencies. First, an exchange can be taken in its literal sense of *money*. If an aircraft cannot depart because the crew has no working hours left, and the passengers who cannot be booked on other flights that day must be put up in a hotel, the costs of solving a problem can escalate quickly. But there are also situations in which offering a hotel to passengers can help to prevent even costlier disruptions from occurring. However, as competition rises, the value of money as 'exchange' decreases.

Second, *capacity*, in all its varieties (seats, crew, planes, slots, runways), is seen by KLM staff as an important form of exchange. For example, KLM keeps reserve crews that can be at Schiphol Airport within an hour. If this crew reserve is 'spent' in resolving a disruption and another one occurs, then a crew must be found among those who are not on standby. Such a change means a compensation in the planned schedule, which itself can create new disruptions elsewhere in the crew schedule.

Third, *anticipation, knowledge and experience* are non-material forms of exchange. Front Office staff emphasized the importance of what they called 'proactive work', by which they meant the anticipation of delays and disruptions, which can create more room for acting. By looking at news and weather reports, OCs learn where to expect certain kinds of problem. They know from experience that using an aircraft scheduled for maintenance as a substitute for a damaged one can lead to subsequent disruptions. And cancelling flights, they argue, is sometimes necessary in order to prevent the effects of delays from snowballing into subsequent days.

Fourth, what sort of information KLM staff have at their disposal in making decisions depends in part on the *communications and information technologies* they use. The OCs in the Front Office use conference telephoning, Internet and satellite telephoning simultaneously, which enables them to reach a pilot anywhere in the world, as well as older means of communication, such as walkie-talkies and telex, that are used to connect different parts of the airport. Closely linked to these communications technologies are information technologies such as the Traffic Flow Management Decision Support Tool and the flight information, check-in and reservation systems. By using these information technologies, controllers are able to monitor actions and events that are spatially distributed, thereby creating a 'virtual situation' on which they are able to act by connecting and disconnecting distanced actions and events. Virtual situations not only provide an overview, they also provide a means of acting on all parts of them. Knowing that a plane which is scheduled for maintenance can be used as a reserve increases the number of possible actions that can be taken in case of further contingencies.

A fifth currency of exchange is *risk*. My field interviews made clear again and again that taking risks should never be considered as exchange. According to the KLM staff, the safety of passengers and crew is not negotiable, a claim that is crucial for ensuring the trust of passengers. In reality it is an implicit form of exchange that is used occasionally, for example in relation to the maintenance schedules of airlines. Defects in an aeroplane that are considered to be a safety risk have to be repaired within 24 hours. Other repairs can wait longer.

A final example of exchange is *authority*. Whether an employee or division can solve a disruption depends on whether they have the authority to make the decisions required to resolve it. While the Duty Manager Operations stands at the top of the hierarchical pyramid, it is only in extraordinary situations, such as accidents or hijackings that this amount of centralization of authority is needed. In KLM's daily operations, the OCs make the final decisions on delaying or cancelling flights. The people in the Hub Control Centre in turn have the authority to decide on prioritizing the platform handling of individual planes above others, but only if the delay this causes does not exceed fifteen minutes after scheduled departure time. Sometimes the use of authority involves possessing *sympathy* for others working in difficult situations. One Senior OC said he realized what it meant for people on the shop floor when plans and schedules were changed at the last minute:

> The way we approach them, the tone of voice we use is important to get the job done. When you give them a call afterwards to tell them everything worked out well, they will be prepared to go the extra mile again next time.
>
> (Interview with Senior OC)

Conclusion: balancing acts

The concept of 'exchange' adds yet another dimension to contextualized accounts of human action in distributed situations. Exchange is relational. If there is little time, then money, capacity, experience, information and communication technologies, taking risks, authority and sympathy all become more important. This explains, for example, why planning weeks and months ahead saves airlines money, and why under conditions of time pressure, exchanging money may be the only way to solve a problem.

Yet, during my field work in 2007, it became clear to me that, contrary to what one would expect, time pressure is not ubiquitous on the platform, departure hall and control centres. The custom of treating the scheduled departure time as a rigid factor for each operation in the network does not reflect what occurs in practice. It is better seen as an ongoing process of negotiation that finally results in the departure of a flight, be it on time or not. In this negotiation process, choices are constantly made at different levels of the operation. The Team Leader has to decide if he proceeds to load baggage that arrives late on the platform and, thereby, delay the flight. The Duty Area Manager can order the platform personnel to stop handling one aircraft and continue with another that is more urgent. The OC can decide to delay a flight in order to accommodate five business class passengers from a late incoming connecting flight, because the traffic flow support tool indicates that the total future revenues for these passengers are high enough to compensate for the costs of the delay.

In the complex, everyday reality of airline operations, every action is planned and scheduled and thereby subjected to a deadline. Yet, as these deadlines are constantly renegotiated, plans and schedules do not precede the actual state of affairs, but are constitutive of the actual situation. If an operational plan can be seen as the projected allocation of resources in time and space, these allocations are no longer fixed when a plan becomes part of the real-time operations. The people at KLM, whether they work on the platform, in the departure hall or in the control centres, need to know what will happen in the next several hours. This explains their constant need for information that is provided by the many real-time information systems. As one Duty Hub Manager explained: 'We can only steer if we know what is going on on the platform, at the gates. That is why we put an enormous effort in teaching people not to concentrate on their own local processes, but to let us know if things go wrong.' In the KLM control centres, there is a daily struggle of extending the 'planning horizon' as far as possible.

The allocation of delay codes is as important as the constant reconstruction of the planning horizon. Where plans represent a reality that precedes the actual state of affairs, reports and delay codes reconstruct what happened after the fact. The allocation of these codes is also subject to intricate negotiations. One day, because of technical reasons, an intercontinental flight had to return to the airport shortly after it took off. Back at the airport, the Maintenance Department took four hours to repair the aircraft. In the meantime, the freshness requirements of the on-board food expired and it had to be changed. Who was to blame for the delay? As the Duty Area Manager explained: 'Maintenance entered a delay code in the system pointing at the catering. But it was their problem in the first place. So I changed it.' The codes allocate responsibilities and accountabilities, and in doing so translate the time pressure following from a general ideal of on-time performance to a multitude of local pressures.

Following 'time' around at an airport when analyzing the daily operations of a big airline such as Air France-KLM at Schiphol airport does not lead to a world of deadlines that need to be met no matter what. Rather 'time' leads to a practice dominated by balancing acts. Analogous to the locks in the cargo hold of the Boeing 747 Combi that had to be adjusted to accommodate the irregularly sized pallet, planned reality must constantly be reshaped to fit the actual flow of events.

Thus, controlling complex aviation networks requires the ability to cope with an ongoing tension between plan and reality, between predictability and contingency. Information and communications technologies, visualized in the countless screens that can be seen in airports, are used extensively to anticipate contingencies, to create an overview of many more or less distant events, and to act at a distance whereby local and central knowledge and experience merge. Summing up, my field work at KLM shows that performing the flight schedules is only possible by anticipating and reacting to changing and emerging situations that have been combined into a 'virtual situation'.

As virtual situations are created in control centres such as KLM's Hub Control Centre and Operations Control Centre, complexity can be both a source of problems and of solutions. It is precisely because the OC Europe connects two events that would not otherwise be related (a damaged plane in Barcelona and a plane scheduled for maintenance) that a problem can be solved, which, in turn, is only possible if no new problems emerge by doing so. The relational character of exchange explains why complexity in aviation does not just lead to a gridlock in increasingly intricate webs of interrelated events that are increasingly vulnerable to small disturbances. On the contrary, the reconstruction of airport time suggests that the daily operations of an airline can only be controlled if there is enough overview to generate the exchange that is needed to solve emerging problems.

Acknowledgements

I wish to thank the employees of KLM who showed me how they work on time, and Daan Nijland of the KLM Operations Control Centre and Richard Reijn of the KLM Hub Control Centre for their time and hospitality.

Notes

1 KLM, or Royal Dutch Airlines, merged with Air France in 2004, making it the largest airline in Europe and the third largest in the world. In 2004, KLM employed 34,529 people and maintained a fleet of 188 aircrafts.
2 The field work was carried out in 2000 and in 2007.
3 The phrase 'minus last minute change' means the total number of passengers minus one at the last moment.
4 Other international IATA delay codes include one for carrying too much hand luggage at the gate (code 10), a defective aircraft (code 41) and a shortage of crew (code 64).
5 The three most important information systems are FIRDA, CORDA and CODECO. FIRDA, an abbreviation for Flight Information Royal Dutch Airlines, is a real-time flight information system. CORDA stands for Computerized Reservations Royal Dutch Airlines. CODECO is the companies check-in system. Both FIRDA and CORDA are mainframe applications that have been in use since the late 1960s.
6 Universal Coordinated Time, the standard time in aviation, is maintained with atomic clocks located around the world. It replaced Greenwich Mean Time in 1928. However, it is sometimes still referred to colloquially as Greenwich Mean Time (GMT).
7 The feedback loops in the KLM flow chart are characteristic of airline disruption management approaches generally. Disruptions in day-to-day operations are analyzed using mathematical modelling and computer algorithms. For examples of this style of reasoning, see Yu (1998) and Yu and Qi (2004).
8 Although the Control Centre's Front Office is brand new, the principles that underlie it go back to the nineteenth century. See Beniger (1986) and Yates (1989) for historical accounts of railway control rooms.
9 There are, however, critical situations that demand the operation to be stopped, as became clear on 11 September 2001, when US air space was closed, and no commercial flights were allowed for several days.
10 This way of doing things is not contested by Operations Control Centre staff. They recognize that in daily practice there is room to deviate from the rules and protocols if the circumstances require it. For one of the Duty Manager Operations, 'rules serve as a guideline, but at the same time, we as sensible men will differ from the rules if necessary' (interview with Duty Manager Operations).

Bibliography

Adey, P. (2007) ' "May I have your attention": airport geographies of spectatorship, position, and (im)mobility', *Environment and Planning D: Society and Space*, 25(3): 515–36.

Augé, M. (1995) *Non-places. Introduction to an anthropology of supermodernity*, London: Verso.

Beniger, J. (1986) The control revolution: technological and economic origins of the information society, Cambridge, Massachusetts: Harvard University Press.

Castells, M. (1996) The rise of the network society, Oxford: Blackwell.

Ciborra, C. U. (1999) 'Notes on improvisation and time in organizations', *Accounting, management and information technology*, 9: 77–94.

—— (ed.) (2000) *From control to drift: the dynamics of corporate information infrastructures*, Oxford: Oxford University Press.

Crossan, M. M. (1998) 'Variations on a theme – improvisation in action', *Organization science: a journal of the Institute of Management Sciences*, 9(5): 593–9.

Derogée, E. (2006) *Non-stop Schiphol*, Luchthaven Schiphol: Contrail Publishers.

Hamilton, A. (2000) 'The art of improvisation and the aesthetics of imperfection', *The British Journal of Aesthetics*, 40(1): 165–85.

Heath, C. and Luff, P. (2000) *Technology in action*, Cambridge: Cambridge University Press.

Kesselring, S. (2006) 'Global Transfer Points. International Airports and the Future of Cities and Regions'. Paper presented at the conference 'Air time-spaces. New methods for researching mobilities', 29–30 September 2006, Lancaster University.

La Porte, T. (1988) 'The United States air traffic system: increasing reliability in the midst of growth', in R. Mayntz and T. P. Hughes (eds.) *The development of large technical systems*, Frankfurt am Main: Campus Verlag: 215–44.

Misa, T. J., Brey, P. and Feenberg, A. (eds.) (2003) *Modernity and Technology*. Cambridge, Massachusetts: The MIT Press.

Moorman, C. and Miner, A. S. (1998) 'Organizational improvisation and organizational memory', *The Academy of Management Review,* 23(4): 698–723.

Orlikowski, W. J. (1996) 'Improvising organizational transformation over time: a situated change perspective', *Information systems research: a journal of the Institute of Management Sciences,* 7(1): 63–92.

Pascoe, D. (2001) *Airspaces*, London: Reaktion.

Peters, P. F. (2006). *Time, Innovation and Mobilities*. London: Routledge.

Suchman, L. A. (1987) *Plans and situated actions: the problem of human-machine communication*, Cambridge: Cambridge University Press.

—— (1993) 'Technologies of accountability: of lizards and aeroplanes', in G. Button (ed.) *Technology in working order: studies of work, interaction, and technology*, London: Routledge: 113–26.

—— (1997) 'Centres of coordination: a case and some themes', in L. B. Resnick (ed.) *Discourse, tools, and reasoning: essays on situated cognition*, Berlin: Springer.

Yates, J. (1989) *Control through communication: the rise of system in American management*, Baltimore: The Johns Hopkins University Press.

Yu, G. (ed.) (1998) *Operations research in the airline industry*. Boston: Kluwer Academic.

—— and Qi, X. (2004) *Disruption management: framework, models and applications*. Singapore: World Scientific.

9   A life in corridors[1]

Social perspectives on aeromobility and work in knowledge organizations

Claus Lassen

Introduction

Flying is one of humankind's oldest dreams. In the beginning of 'the age of flying' it fulfilled a dream so long held by scientists, philosophers, artists and poets. In the 1950s and 1960s flying was associated with glossiness and luxury (Blatner 2003: 87). But today flying is a fundamental element in economic and cultural globalization (Graham 1995). There are four million air passengers each day, 1.6 billion air journeys each year, and at any time there are 300,000 passengers in flight above the USA (Urry 2003a: 157). Within the EU alone, the total number of passengers transported by air rose to more than 700 million in 2005 (European Commission 2007). National statistics show that air traffic between Denmark and foreign countries increased markedly within the past decade in both flights and passengers. The number of passengers on international flights has increased by 76 per cent and international take-offs and landings by 98 per cent between 1990 and 2006 (Danmarks Statistik 2006).

Increased aeromobility in the modern society is related to the building of monumental airports of glass and steel, designed by celebrity architects and supported through mega infrastructural projects (Flyvbjerg *et al.* 2003) of motorways and high-speed trains, gigantic aeroplanes and flights greatly cheaper than surface travel. These are icons of the new global order (see Urry (Chapter 1), this volume). Moreover, aeromobility also delivers material support to various new patterns of identity and lifestyle and generates many so far unexplored social consequences (Sheller and Urry 2006: 208; but see the chapters in this volume). However, the relations between modern society and increased flying in relation to work, family, love and leisure are overlooked in the social sciences, either by accident or design (Pascoe 2001: 7).

One important element of the increase in aeromobility is work-related travel taking place on a global scale. Figures from the World Tourism Organization (2005) show that 19 per cent of all international travel each year has a work-related purpose. And this type of travel has more than doubled between 1990 and 2001. A survey from the Danish Transport Council shows that 40 per cent of all international air travel from airports in Denmark has a work-related purpose (Transportrådet 2001: 4). The increase in international work-related travel by plane seems especially associated within knowledge organizations[2] (see Lassen (2006) and Høyer and Næss (2001) for further elaboration) as an important part of the so-called 'new knowledge economy', based on networking, flexibility and mobility (Castells 1996: 77).

This chapter therefore explores international work-related travel as an important part of the increase in aeromobility, especially focusing upon two Danish international knowledge organizations. It argues that work in international knowledge organizations cannot be understood separate from aeromobility organized through a large material system of corridors. The corridors not only materially support working life in knowledge organizations but also function as a logic of action. It is a selection mechanism, which picks and chooses so that the traveller is distributed in accordance with the logic of the corridor – a logic anchored in the 'space of flows'. However, the chapter also shows that knowledge workers approach such systems of corridors differently and produce varied patterns of aeromobility. Therefore, the social consequences of 'a life in corridors' are experienced differently. For some knowledge workers, aeromobility means new opportunities to network, to combine work and pleasure, to develop a cosmopolitan identity, to play in new places; but for others, it involves a great deal of frustration and ambivalence in relation to coping with work, family, leisure, localities, as well as belonging in between the global and local. In conclusion, the chapter therefore critiques the idea of the 'happy cosmopolitan' living a carefree life on the move for future research on the complex connections of aeromobility and work.

Exploring work and aeromobility

Theoretically, the chapter is placed within the new 'paradigm of mobilities' (Beckmann 2001; Kaufmann 2002; Kesselring 2006; Lassen 2005; Sheller and Urry 2006; Urry 2000, 2002). It seeks to understand the sociology of international work-related travel and aeromobility. Raising such questions as: 'what are the mechanisms and patterns of meaning associated with work-related travel by plane?', the overall approach is not to focus on air transport in the specific types and forms of transport but instead upon more general processes of 'aeromobility'.[3] As a consequence the paper addresses the social bases as well as the social consequences of aeromobility and international work-related travel.

The study of work and aeromobility in knowledge organizations rests on a model of understanding that covers mobility, identity and work. This model of understanding especially examines how individuals create and cope with strategies that give meaning to them in everyday life and expresses a meaningful and manageable handling of external demands and internal intentions (Lassen and Jensen 2004: 252). Moreover, the study not only focuses on the actual mobility itself, but also on the potential mobility of people, and how potential mobility is transformed into different forms of physical, virtual and social mobilities (Høyer 2000; Kaufmann 2002; Kesselring 2006). Knowledge workers who face a demand of aeromobility develop different coping strategies that produce meanings for them (Bærenholdt 2001). With a starting point in a 'critical realist' position, the mobility strategy of individuals and their practice is considered to be in a causal/material relationship with its surroundings (Bhaskar 1975; Sayer 2000). The analytical approach therefore focuses on both the production of meaning *and* causality. Drawing on such theoretical approach the analyses are empirical, primarily based on a multiple-case study (Flyvbjerg 2001). The case study involves Aalborg University and Hewlett-Packard Denmark,[4] and the method involves, first, a web-based questionnaire sent to all employees in the two cases (a total of 1,800 employees); and second, face-to-face interviews with selected employees in the two cases (eleven employees).

In the following I will go through some of the main results. First, I identify the main characteristics of work and aeromobility in the two cases. Second, some of the principal rationalities for international travelling are analyzed. Third, I show how employees variably approach work and aeromobility, which means that the social consequences are differently experienced. Finally, there is a brief conclusion.

The nature of corridors

The case study shows that international work-related travel and aeromobility are very important components for the knowledge workers at Hewlett-Packard and Aalborg University (see Lassen *et al.* (2006) for further details). A general outline related to the 'narratives' is that the employees describe international work-related travel as a 'practice which takes place in different corridors', for example: in airports, in aeroplanes, on motorways, in hotels, in offices and even the movement of the aeroplane takes place in corridors (Blatner 2003: 63). The corridors have similarities with what Castells (1996) terms 'space of flows' and Kvaløy (1973) terms 'system of channels' as ways of organizing contemporary social practices. The international work-related travelling evolves in high-speed spaces, where the employees are moving on the way to the next meeting, next hotel, next bar and next country (see Lassen (2006), for a further description). When the employees travel they contribute to the construction of corridors through spatial practice, but their own cognitive experience and logic of action are also influenced by movement through the corridors. Thus, the reference here is of a spatial organization, where the corridors function as a selection mechanism, which picks and chooses so that the traveller is distributed in accordance with the logic of the corridor – a logic anchored in 'space of flows'. The corridors deliver, like the space of flows, both a logic of action and a material spatial origination of social practice (Castells 1996: 406). It is often difficult to distinguish between the different places within corridors, which means that the movement through the corridors is experienced as monotonous. One male employee from Hewlett-Packard, who often travels long distances to different countries, exemplified this experience of monotony. Sometimes, he explained, it can be difficult to determine where you are exactly, because offices, hotel rooms and airports look the same across the world:

> One of the disadvantages with travelling for a company is that our offices look the same, so you can wake up in a hotel room and say to yourself: Where the hell am I? Is this Denmark, Norway or Sweden? You actually don't know. It's very impersonal, isn't it?
>
> (51-year-old male manager at Hewlett-Packard)

This illustrates how the corridors are like 'non-places'. The corridors create a material frame for people moving from one non-place to another. Such places are non-historical, non-relational and have no self-identity and, furthermore, they are shaped by the 'space of flows'. Non-places are, for example, represented by the airport, motorway café and shopping centre. According to Augé, non-places are opposite to anthropological places. The anthropological place is, for example, the historical city centre or an old fishing village on the coast, and is characterized by relationality, history and self-identity (Augé 1995: 77). 'Non-places' are characterized by what Castells terms 'the architecture of nudity' (1996: 450), in which the forms are so neutral, so pure and so diaphanous that they do not pretend to say anything. The airport exemplifies a non-place that includes these types of architecture (Castells 1996: 451). According to Jensen, the concept of non-place implies the risk that we only picture the downside and the cultural degradations emerging from 'non-places' (2007: 7). Jensen therefore argues that we need to see potentials for new public domains, and he states that we can benefit from seeing spaces of flow as potential sites of resistance, meaningful social communication and interaction, instead of just generic non-places (2007: 21). In relation to the places of the corridors, we therefore need a more differentiated understanding, which I show in the following sections. The corridors not only consist of non-places, but also of places that offer space for meaningful social communication and interactions in order to maintain and develop network relations, identities, leisure and tourist activities, professional skills and escape from the time-pressure of everyday life etc.

The aeroplane as a global bus

The aeroplane is fundamental in the construction of the corridors. For the knowledge worker of today flying is not a novelty, the aeroplane is just a simple means of transport in order to get to a requested location. A male frequent flyer at Hewlett-Packard describes how his relationship to air travel has changed since he first entered a plane. He was then rather nervous, but today the plane can fly in air turbulence and he will not notice it, because travel has become a routine and a common practice. Aeroplane travel is no longer considered a special activity, because the meaning of international air travel has transformed into having the same meaning and function for the cosmo-politan professional knowledge worker as the bus trip has for the industrial worker. This travelling experience is fundamentally different from the attrac-tion and mystery travelling abroad had historically, when for example an explorer went out in the world and came home with marvellous travelogues about the mythically foreign countries he has visited (see De Botton 2002). A fairytale, in Bauman's words, loses its meaning when the whole life becomes a number of fairytales, and travel, in the same way, loses its aura when it becomes everyday life (1999: 96). In the corridors there is no alternative to the plane. In general, the employees express how they very seldom consider alternative modes of transport because aeroplanes are for them the only realistic possibility. As one employee from Hewlett-Packard says: 'I never consider other options than flying, actually I only consider which air company I want to travel with'. Travelling by aeroplane in corridors is founded in a rationality of 'clock time' (Jensen 2001: 53), which means that the employees constantly try to go as efficiently and as fast as possible between aeroplanes, taxis, hotels and places of work. This is considered natural in the corridors and is unquestioned. The corridors offer no room for alternative ways of action. High-speed air movement in the corridors contributes to why employees consider the trips monotonous and boring, because one cannot see much through the plane window, and it leaves no room for experiencing and sensing the places and cultures the employees move through.

The nodes of the corridors

The airport is one of the central non-places in the corridors. It is the system of airports that is key to many global processes, permitting travellers to encounter many people and places from around the world, face to face (Gogia 2003). Airports have been described as 'mono cultural zero friction enclaves' (Hajer 1999). These spaces function in a disciplined way, as a machine for movement designed to move people smoothly between one form of transport and another. However, the employees also experience the airport as a break in their mobility, because it is also related to waiting time and slowness (see Fuller (Chapter 3), this volume). There is a narrowing of movement, slowing down the hectic and high-speed movement through the corridors. The airport includes a number of places of friction, for example the baggage reclaim and waiting areas. In the airport, high-speed movements encounter places of slowness and immobility:

> I only become more tired, I try not to stress because I think that you can't do anything about it anyway; if the plane is delayed then it is delayed. I am very good at arriving last minute and that probably stresses me unnecessarily. It is like chasing the clock. It is hard to chase the clock all the time, then there is one thing you have to catch, then there is another

thing that you will have to catch. You have to catch an aeroplane here, catch a meeting there, and so on, and so on.

(42-year-old female consultant at Hewlett-Packard)

Sometimes high-speed movement slows down, and this can be stressful for some travellers because they cannot follow the temporal logic of the corridors. Suddenly it is not possible to chase the clock. This view on airports as time-consuming is increased the more the employees are socialized into the logic of the corridors. In general, the employees experience how friction has increased after 9/11, because of more security. The temporality and spatiality of the corridors mean that, for some, a stay in airports is stressful because of the things the employees have to be in time for. Others of the employees have, as Castells points out, realized that when they are in the middle of the 'space of flows', they are, literally, in the hands of the airport and the air companies, and there is no way of escaping (1996: 451). This means that the stay in the airport is used as a working place, as they also use other of the non-places in the corridors as working places (such as the hotel room). Work is carried out on their laptop computer, which they always bring with them. They make a number of calls to colleagues, network relations, friends and family while they are waiting to get on the plane, which illustrates that the airport is also a space of virtual socializing. In general, the airports are therefore places of the 'boring everyday routine', not only for the thousands of workers located within airports-cities (Sheller and Urry 2006: 216) but also for the users of the corridors. Other very important places in the corridors are the international hotels, open-plan offices and the conference rooms. At Aalborg University, central places are the conference and meeting rooms as well as the hotels. At Hewlett-Packard, the employees mostly spend their time on an international trip in open-plan offices and hotel rooms, and they often work many hours during the day, which increases the feeling of 'a life in corridors'.

In summary, the knowledge workers consider high-speed movement through the corridors monotonous, stressful, boring and non-human. However, this is only partly the story of the life in corridors. International work-related travelling is related to a number of other meanings and rationalities in relation to work, family and tourism that makes it attractive and necessary for employees to step into the systems of corridors. I will explore these in the following section.

Why step into the corridors?

First of all, a core of obligations can be identified where the employees, through culturally embedded expectations, are expected to be present in relation to specific events, places and people. In both cases a core of obligations is identified that is very difficult for the employees to decline if they want to keep their present job. The employees describe the physical meeting 'face-to-face' and 'co-presence' as important in relation to being a part of the 'network driven workplace'. A face-to-face meeting with customers in order to close a deal or a conference dinner can be important for the purpose of managing the job. In this relation, co-presence offers the possibility of establishing intimacy and trust (Urry 2002; Boden and Molotch 1994). Attending a meeting, a course, a conference, etc., has much more to it than the act itself. Both formal and informal agendas exist, where networking in the corridor or informal 'meetings' over a cup of coffee are very important practices. Such physical face-to-face meetings and co-presence offer something that cannot be gained through virtual communication, such as eye contact, body language, small talk, socializing and so on (Boden and Molotch 1994: 269). As Riain points out, 'face-to-face interaction' remains a critical component of the global workplace, and transnational 'virtual' relationships are constantly supplemented by international travelling (2000: 185–7). As a response to the demands of mobility in the knowledge organizations, employees are dependent on networks because they give access to knowledge and resources. Networks open doors when it comes to everyday problems at work and strengthen one's personal career. These networks can often be 'international', and this heightens the significance of networking tools. The ability to create what I term *acquaintance* and *friendship* is important in developing networks. It is important to know the right persons and to be well-known among them (Wellman and Haythornthwaite 1999: xviii). To develop acquaintances and friendships, the employees from time to time need to travel to places for '*organized fortuitousness*' (see Lassen 2005: chapter 4). Places for organized fortuitousness are conferences, courses, workshops, business meetings and so on. Places for organized fortuitousness cannot be understood as a non-place but as very meaningful. To be present at places where accidental meetings are likely to occur is necessary to develop and keep acquaintances and friends. Here the international work travel through corridors plays a very important part, because it makes it possible for employees to be at the right place at the right time to socialize with others and to share moments of co-presence and face-to-face communication (Urry 2002; Wittel 2001). Aeromobility can be significant in developing and extending networked connections.

Second, international mobility is important because it offers material support to a 'cosmopolitan' identity. Hence, a connection can be identified between the forms of mobility that are practised by the employees in their networked everyday lives and the question about social identity, e.g. ways of living in relation to choosing a means of transport. Among other things, employees use 'identity accessories' in their work of constructions, not only locations and places, but also the movement between these places and locations. For a hyper-mobile 'way of living' or 'way of working', 'consumption of distance' becomes a fundamental element, and there is a connection between this type of consumption and particular human lifestyles and identities (Whitelegg 1997: 59). A cosmopolitan identity puts the spotlight on the prevalence of this new kind of cosmopolitanism or post-national situation, where focus is the global civil society as the frame of reference for solidarity and identity (see Castells 1996; Habermas 2001; Stevenson 2003; Urry 2003a; Beck 2002). There are increasing streams of cosmopolitanism derived from television, aeroplane trips, mobile phones and Internet connections. This transforms the relationship between co-presence and mediated social relations, between proximity and remoteness and between local and global (Harry 2000 in Urry 2003a: 138). In this way, Urry states that: 'Cosmopolitan fluidity thus involves the capacity to live simultaneously in both the global and the local, in the distant and the proximate, in the universal and the particular' (Urry 2003a: 137). This kind of cosmopolitanism will, to an increasing extent, affect civil society and redefine the conditions under which social actors are gathered, organized and mobilized (Urry 2003a). According to Beck (2002) these cosmopolitan streams originate from the transformation processes that reduce the importance of national borders, support time-space compression and increase the international network relations between national societies. This is especially exemplified by the globalization of the economy. However, there are other contributing elements, in that humans to an increasing extent trade internationally, work internationally, love internationally, marry internationally, do research internationally, grow up and are educated internationally, and finally think internationally. As it appears from the above mentioned references, a cosmopolitan framework of understanding therefore changes the classical understanding of identity as unchangeable to an understanding of a new hybrid and global identity formation process, where social actors are seen as having both roots and wings (Zachary 2000: 49).

Finally, international mobility offers employees the possibility of 'escaping' from the constraints and monotony of corridors when they are 'on the move'. The chance of experiencing the *genius loci* and the local culture outside the corridors' monotonous spaces means that international work-related travel is also attractive. The employees move through corridors linked together by nodes of hotels, airports, companies, etc. But the employees do not only work when they are abroad. Sometimes they leave the corridors in order to visit tourist places, to go sightseeing or to experience historical localities, alone or in company with colleagues, friends and family Or sometimes they visit friends or family abroad. These types of activity usually bring excitement and distraction to monotonous, work-related aeromobility; it is a way of escaping from the 'life in the corridors' (see also Rojek 1993). This 'escape' from the corridors seems to bring curiosity into the trip, when the employees leave the corridors behind and enter into a more colourful and attractive world.

Different practices in relation to mixing work and pleasure can be identified.[5] One way of combining work and pleasure activities can be termed 'day tripper activities'. In both of the two cases, the employees carry out tripper activities when they are on international work-related trips. One employee from Hewlett-Packard describes how she often travels to places that she would never have gone to as a 'private' person. She therefore takes short trips in the surrounding area of the place that she is visiting. It gives her a chance to explore the place and it brings excitement into her journey. Sometimes she leaves the corridors and colleagues accompany her. In one place, the employees go shopping, in another place they go sightseeing. A female employee from Hewlett-Packard often organizes different events in relation to meetings. If for instance a meeting takes place in Lyon, they perhaps organize wine tasting. If the meeting is in Amsterdam, they might go for a boat trip on the canals. From time to time, the employees use the opportunity to collect objects that are symbolic of the places they visit. Symbols that will work as memories of the place and that can demonstrate to others that they are widely travelled. Such souvenirs are often a photo, a T-shirt with a well-known tourist site on the front, or a special item characteristic of the attraction, city, nation or part of the world they have visited.

Another practice can be termed 'tourist activities'. Some of the employees occasionally use the opportunity to extend the normal working trip by a few days, or maybe they use it as a launch pad for a vacation. Therefore, they extend their travel by some extra days or start their holidays at the end of an international working trip. Employees sometimes travel to interesting places and want to explore these places more than is possible within the ordinary working period abroad. This could for example be to have a holiday after participating in conferences in, say, Greece or Australia. This practice also offers opportunities of collecting symbols, but moreover it gives the employee a chance of escaping from the time pressure within the working life, the family and not least the corridor.

Furthermore, international work travels involve a practice that I term 'family activities'. This includes both 'bringing and visiting family and friends' abroad. It happens that employees, especially from the university, bring their partners, children or friends along with them. An example is a male professor who is doing research on international relations and for whom travelling has been a normal element in his and his wife's life. Often his wife meets with him at the destination at the end of a working trip, and together they have a holiday. Others bring their family on the full-length international trip or use it to visit friends and family who live abroad.

Approaching the corridors differently

Mobility is not homogeneous but is produced differently inside the same field. In the following I argue that there are three different ways of approaching the corridors and dealing with work, family and aeromobility, namely career strategy, juggling strategy and family strategy.[6]

The first is *career strategy*. These employees often live a single life or have a partner who functions in the background and takes care of all everyday aspects of family life (Christensen 1988). Life is work, and aeromobility is a fundamental and necessary element of the working life. The employees function as 'cosmopolitans', living life in different countries and international networks. These are what Bauman (1999) terms 'cosmopolitan tourists' and Riain (2000) terms 'software cowboys'. A high level of yearly international travel and an international orientation are core elements in the self-conception and identity. The following exemplifies the career strategy:

> It is enormously important, it depends on how one means the international. I am not thinking much in national terms and when I open the newspaper I am not reading much about the hospital sector or what it might be, I go straight to the international news; I am focused on the international. I don't consider myself Danish through and through either, if one can put it that way, even though I am so by birth, but I think that I have more of an international self-conception than a national self-conception.
>
> (40-year-old female Associate Professor from Aalborg University)

The employees practise a de-centred handling of mobility (see Kesselring 2006: 272), where the home is less clear than for the other strategies. They use virtual technologies such as the mobile phone and laptop computers as tools to stay in touch with friends and relatives when they are on the move. Employees with this strategy also have spare-time interests that are not place-bound, such as writing a book or running! International air travel enables a life without limitations of place-bound activities, while virtual communications offer the possibility to be away and present at the same moment (Urry 2003b). For the members of the career strategy, life in corridors seems unproblematic, because it offers a lot of possibilities (described above), and there are no conflicts with other elements of everyday life.

The second strategy is the *juggling strategy*. This means that both family and work have high priority for the employees inside this strategy without the opting for one in favour of the other. The strategy is based on a 'glocal'[7] (Urry 2003a) identity, where the employees try to combine and juggle their international working life and a locally based family life. These employees do not want to prioritize international working life on the one hand and family and local relations on the other. They construct their identity as a cosmopolitan and as a family person. The working life therefore puts family life under pressure, and the employee tries to compensate by diary planning and 'quality time' (see also Hochschild (1997)). Everyday life is like a line dance, where the employees constantly try to cope with work, family and international travel, without falling down. The employees in this strategy use a centred mobility strategy (Kesselring 2006: 271), where they travel back and forth between home and abroad in lined patterns of movement through the corridors. The juggling strategy expresses the most ambivalent and stress-causing strategy among the employees, because it is difficult to make the various elements of everyday life fit together in a non-stressful way. One exemplification of this can be seen below:

> But it is probably to a larger extent the change between being abroad, living at a hotel, where you only have to think about work. When you come home,

you suddenly have to wash up, do shopping, see to it that there is food in the fridge and things like that. I think that no matter whether you have children, this will always be a dilemma. It is difficult to make that switch-over between being abroad where you are being served. You can just ask for what you want . . . this change between the two worlds, I find hard, really hard. It is sometimes like that when you come home late, there you are dropped down, and then you have to get up next morning and follow the child to school. This I think can be hard, especially if you got home late.

(42-year-old female consultant from
Hewlett-Packard)

It is difficult, both practically and mentally, to shift between staying at the Hilton in London in the morning and participating in a school meeting in the afternoon in Copenhagen. The employees with the juggling strategy often experience longing to be in another place. When they are abroad, they long for the children and when they are at home, they long for more time to themselves – away from everyday problems. For this group of employees it is a clash between two different worlds, two different types of rationality and two different types of place logic: the logic of space of flows and the logic of space of places (Castells 1996). Research showed how these employees are not very reflexive about why they often have a feeling of stress and ambivalence in everyday life.

The third strategy is the *family strategy*. The members of this strategy give a very high priority to family life. It is a clear priority to place family life before work, and aeromobility is planned with consideration for the routines of family life. The identity is also 'glocal' (Urry 2003a), but where the local relations take precedence over global elements. International travelling is seen as an important, exciting and exotic part of the job but *only* if it fits in with other obligations of family life. One male employee from Hewlett-Packard exemplifies this strategy:

Of course, one is interested in doing some of the things one does, instead of sitting around twiddling one's thumbs, so if one hasn't got other things to do in one's nearest geography, well, then you travel for it, not more not less. I am not one of those who travel far for assignments. There are some of my colleagues who have been to South Africa or Saudi Arabia, Asia, the USA, and South America, and of course it has a lot to say how your domestic situation is. If one is a bachelor and hasn't got children it's a lot easier to go away for a week or a month or whatever the need is. For example, some of my colleagues have been stationed for two or three years in the USA. As to myself, I try not to travel too much and preferably not too far.

(41-year-old male consultant from
Hewlett-Packard)

These employees seem to take reflexive and strategic decisions, balancing freedom and compulsion in everyday life organization of mobility. Also here the employees practise a centred strategy of mobility (Kesselring 2006: 272), which can be less stressful than the juggling strategy, because the family is the centre and priority of action in life. Of course international travel also causes problems in family life but on a much lower level than the juggling strategy. Family strategy also seems to be the most important 'brake of aeromobility' (see also Gustafson (2005; 2006)). Family relations play an important role in the light of how much the employees are willing to travel internationally. The employees practising a family strategy are internationally oriented and like to travel internationally but they emphasize the meaning of local relations more than the other strategies. Therefore, international work-related travel is kept to an absolute minimum. Virtual mobility is used to network internationally and to stay in touch with important contacts abroad as a way of substituting the need for co-presence and face-to-face relations.

In relation to the different coping strategies that the staff members use, there is also a gender difference. Generally, the quantitative and qualitative research shows that men travel more than women. Most travel is undertaken by young men between 25 and 30 years old, with no family, whereas those that travel the least are single women, between 30 and 40 years old, with children. Life in the corridors is to a large extent male-dominated. In relation to the social consequences and the gender question, it is worth noticing that those employing the career strategy are largely young men with no children, while those employing the juggling strategy are more likely to be women and people with children. Thus the social consequences of international travel are very differently experienced.

Aeromobility and ambivalence

In this paper I have shown that working in knowledge organizations is lived in corridors organized around aeromobility. I have also shown how the employees approach the corridors differently and I argued that this means that corridors cause different social consequences for employees. The different strategies express different degrees of aeromobility. For the career and family strategy, the social consequences of aeromobility are experienced less dramatically than for the juggling strategy. The first two mentioned strategies consider, in different ways, aeromobility as something relatively positive, contrary to the strategy of juggling where aeromobility is connected with stress and conflicts in everyday life.

This raises a number of questions in relation to the idea of the happy cosmopolitan traveller. Bauman (1999) describes the *cosmopolitan tourist* who, opposite to the *vagabond*, lives a happy and careless life 'on the move'. The tourist is, according to Bauman, the globally mobile, which includes an extraterritorial group of global businessmen, global culture managers or global academics. For these travellers space has lost its constraining quality and is easily traversed in both its 'real' and 'virtual' renditions. The tourist knows that they will not stay for long when they arrive but, unlike the vagabond, they live this extraterritoriality as a privilege, as independence, as the right to be free, free to choose. Opposite to this are vagabonds, who do not know how long they will stay, and more often than not it will not be for them to decide when their stay will come to an end. The vagabond is a pilgrim without a destination; a nomad without an itinerary. The tourists travel because they want to; the vagabonds because they have no other bearable choice. Bauman's metaphor of tourist/vagabond also represents a distinction between mobile/immobile, elite/mass and privileged/unprivileged.

In this chapter I have been critical of the 'happy' and privileged cosmopolitan traveller. Of course a group of happy and careless cosmopolitan tourists do exist, who go and stay as they like, as with the career strategy. But there are also knowledge workers with family and children who do not travel in this unproblematic way (see also Gustafson (2006)). Many world travellers with family face ambivalence each day. These problems arise because people are caught between the logic of the corridors on the one hand, and the logic of family and local life on the other hand. Employees often experience that the separation between travel and family gives them feelings of guilt and longings when they are abroad and, in practical terms, makes it difficult to cope with everyday activities because it involves so many different scales and locations in time and space. This represents a clash between the logic of flows and the logic of places (Castells 1996). As Castells points out, unless cultural, political and physical bridges are deliberately built between these two forms of space, we may be heading towards life in parallel universes whose times cannot meet because they are warped into different dimensions of a social hyperspace (1996: 359). In addition it is also important to emphasize that, for knowledge workers in general, space has not lost its constraining quality. Spaces, as for example places of organized fortuitousness, are very important and meaningful in the network-driven workplace, and it is necessary for knowledge workers to travel to such places from time to time to socialize and interact with others. For some people these circumstances contribute to a split in their everyday life between workplaces and places for family life, because there are large geographical distances between them.

Furthermore, one hypothesis is that the career strategy is the most common strategy in the knowledge economy and, therefore, it is likely that the problem with ambivalence and stress in relation to aeromobility is more widespread in the new labour market than my analysis above indicates by identifying three different coping strategies. Another hypothesis that can be established is that, in relation to the staff members' different ways of coping with international work travel, there is gender stratification, where young men with no family experience the trip as unproblematic, but women with children find it more difficult to handle the gap between the different places, logics and temporalities.

This raises a number of challenges to researchers and politicians. First, theoretically there is a need for developing new metaphors for the cosmopolitan world traveller that cover the negative social consequences that some experience. It is important to consider the world traveller as an individual with wings and roots. Second, this also has political implications, because it is important to create a platform for articulation of the human problems related to long-distance work travel. All this can help knowledge workers to understand better the problems they are dealing with in everyday life. But such discussions should also be addressed to policymakers who design policy, as for example ministries of employment, unions, employers' associations, as well as transport planning and architecture, etc. In general there is a need for much more research on work and aeromobility in order to develop a deeper understanding of the social consequences of such phenomena.

Notes

1   The metaphor is borrowed from Karl G. Høyer, Western Norway Research Institute. He introduced this metaphor at a working seminar held at Aalborg University in 2001.
2   A 'knowledge organization' is defined as an organization that has a high level of either production of knowledge or consumption of knowledge (see Castells (1996) for a more detailed description). I use the term 'knowledge organization' in this paper because it varies from the traditional Fordist organizations and Fordist labour market. Knowledge has, of course, always played an important role in industry, but today we see that units or agents in the new economy fundamentally depend upon their capacity to generate, process and apply efficient knowledge-based information (Castells 1996: 77).
3   The term aeromobility relates to the process of air traffic as a parallel to automobility (Urry 2000: 59). The term aeromobility is inspired by Høyer (2000: 193). Furthermore, aeromobility in this chapter refers to both factual air travel of individuals and their capacity to carry out air-based mobility (Kaufmann 2002: 1). This means that to understand the production of air traffic one may not only study people's factual movement but also their potential to carry out different types of mobility and, in relation to this, understand which mechanisms transform/not-transform potential mobility into actual mobility.
4   There are different arguments to choose in these two specific cases. A survey from the Danish Transport Council (Transportrådet 2001) shows that private-sector employees are mostly travelling in relation to meetings and business purposes, while the employees from the public sector most frequently are travelling to a course or a conference. Furthermore, the organizations are very dissimilar when it comes to core areas of work, organizational structures and management. This means that we are dealing with two very different organizations with the common denominator that their employees are travelling internationally at a high level in relation to work. Hopefully, this will contribute to give the data material empirical depth and thereby give better possibilities to understand international work-related travel and aeromobility as a social practice.
5   See Lassen (2006) for a further elaboration on this subject.
6   The development of career strategy, juggling strategy and family strategy is inspired by an analysis of working life, family life and stress produced by The Danish Society of Engineers (IDA) in 2002; available at ida.dk/Ansat/Ansaettelses vilkaar/Arbejdsmiljoe/Publikationer.htm (accessed 3 December 2007).

7   The term 'glocal' indicates that the global and local are inextricably and irreversibly bound together through a dynamic relationship (Urry 2003a: 84). We can thus talk of a 'glocalized cosmopolitanism' in which 'in the everyday lifestyle choices they make, cosmopolitans need routinely to experience the wider world as touching their local life world, and vice versa' (Tomlinson 1999 in Urry 2003: 137).

Bibliography

Augé, M. (1995) *Non-places: Introduction to an anthropology of supermodernity*, London: Verso.

Bærenholdt, J. O. (2001) 'Territorialitet, mobilitet og mestringsstrategier', in K. Simonsen (ed.) *Praksis, rum og mobilitet*, Roskilde: Roskilde Universitets Forlag.

Bauman, Z. (1999) *Globalisering: De menneskelige konsekvenser*, København: Hans Reitzels Forlag.

Beck, U. (2002) 'The Cosmopolitan Perspective: Sociology of the second age modernity', *British Journal of Sociology*, 51: 79–105.

Beckmann, J. (2001) *Risky Mobility: The filtering of automobility's unintended consequences*, Ph.D. dissertation, Department of Sociology, Copenhagen University.

Bhaskar, R. (1975) *A Realist Theory of Science*, Leeds: Leeds Books.

Boden, D. and Molotch, H. L. (1994) 'The Compulsion of Proximity', in R. Friedland and D. Boden (eds.) *NowHere. Space, time and modernity*, Berkeley: University of California Press.

Blatner, D. (2003) *The Flying Book: Everything you've ever wondered about flying on airlines*, London: Penguin Books.

Castells, M. (1996) *The Information Age: Economy, Society and Culture*, vol. 1: *The Rise of the Network Society*, Oxford: Blackwell.

Christensen, L. R. (1988) *Livsformer i Danmark*, København: Samfundsfagsnyt.

Danmarks Statistik (2006) *Transport 2006*, København: Danmarks Statistik.

De Botton, A. (2002) *The Art of Travel*, London: Penguin Group.

European Commission (2007): *Air transport in the EU25*. Online. Available at epp.eurostat.ec.europa.eu (accessed 1 December 2007).

Flyvbjerg, B. (2001) *Making Social Science Matter: Why social inquiry fails and how it can succeed again*, Cambridge: Cambridge University Press.

—— Bruzelius, N. and Rothengatter, W. (2003) *Megaprojects and risk: an anatomy of ambition*, Cambridge: Cambridge University Press.

Gogia, N. (2003) *Bodies on the move: The Politics and Poetics of Corporeal Mobility*, Toronto: University of Toronto.

Graham, B. (1995) *Geography and Air Transport*, Chichester: Wiley.

Gustafson, P. (2005) *Resor i arbetet: En kartläggning av svenskarnas tränsteresor 1995–2001*, Forskningsrapport nr. 135, Sociologiska Institution, Göteborg: Göteborgs Universitet.

—— (2006) 'Work-related travel, gender and family obligations', *Journal of Work, Employment and Society*, 20: 513–30.

Habermas, J. (2001) *The postnational constellation: political essays*, Cambridge, Massachusetts: MIT Press.

Hajer, M. (1999) 'Zero-Friction Society', *Urban Design Quarterly*, 71: 29–34.

Hochschild, A. R. (1997) *The Time Bind: When Work Becomes Home and Home Becomes Work*, New York: Metropolitan Books.

Høyer, K. G. (2000) *Sustainable Mobility – the concept and its Implications*, Ph.D dissertation, Institute of Environment, Technology and Society, Roskilde University.

—— and Næss, P. (2001) 'Conference tourism: a problem for the environment, as well as for research?', *Journal of Sustainable Tourism*, 9: 541–70.

Jensen, M. (2001) *Tendenser i tiden – en sociologisk analyse af mobilitet, miljø og moderne mennesker*, København: Samfundslitteratur.

Jensen, O. B. (2007) 'Urban Mobility as meaningful Everyday Life practice', paper presented at Annual Meeting of the Association of American Geographers, April 2007, San Francisco.

Kaufmann, V. (2002) *Re-thinking Mobility*, Contemporary Sociology, Hampshire: Ashgate Publishing Limited.

Kesselring, S. (2006) 'Pioneering mobilities: new patterns of movement and motility in a mobile world', *Environment and Planning A*, 38: 269–79.

Kvaløy, S. (1973) 'Økokrise, natur og menneske', *Tidsskrift for Univesitets- og Forskningspolitik*, 8: 119–33.

—— (2005) *Den mobiliserede vidensarbejder: En analyse af internationale arbejdsrejsers sociologi*, Ph.D. dissertation, Institut for samfundsudvikling, Aalborg: Aalborg Universitet.

Lassen, C. (2006) 'Work and Aeromobility', *Environment and Planning A*, 38: 301–12.

—— and Jensen, O. B. (2004) 'Den globale Bus – om arbejdsrejsers betydning i hverdagslivet', in M. H. Jacobsen and J. Tonboe (eds.) *Arbejdssammfund: Den beslaglagte tid og den splittede identitet*, København: Hans Reitzels Forlag.

—— Laugen, B. and Næss, P. (2006) 'Virtual mobility and organizational reality: an examination of mobility needs in knowledge organisations', *Transportation Research Part D*, 11: 459–63.

Pascoe, D. (2001) *Airspaces*, London: Reaktion.

Riain, S. Ó. (2000) 'Net-work for a Living: Irish Software Developers in the Global Workplace', in M. Burawoy, J. A. Blum, S. George, Z. Gille, T. Gowan, L. Haney, M. Klawiter, S. H. Lopez, S. Ó. Riain and M. Thayer (eds.) *Global Ethnography: Forces, Connections and Imaginations in a Postmodern world*, Los Angeles: University of California Press.

Rojek, C. (1993) *Ways of Escape: Modern Transformation in Leisure and Travel*, London: The Macmillan Press.

Sayer, A. (2000) *Realism and Social Science*, London: Sage Publications.

Sheller and Urry (2006) 'The new mobilities paradigm', *Environment and Planning A*, 38: 207–26.

Stevenson, N. (2003) *Cultural citizenship: cosmopolitan questions*, Maidenhead: Open University Press.

Tomlinson, J. (1999) *Globalization and Culture*, Cambridge: Polity.

Transportrådet (2001) *Danskernes flyvaner – en survey*, Rapport nr. 01–02, København: Transportrådet.

Urry, J. (2000) *Sociology beyond Societies: Mobilities for the twenty-first century*, London: Routledge.

—— (2002) 'Mobility and Proximity', *Sociology*, 36: 255–74.

—— (2003a) *Global Complexity*, Oxford: Polity.

—— (2003b) 'Social networks, travel and talk', *British Journal of Sociology*, 54: 155–75.

Wellman, B. and Haythornthwaite, C. (eds.) (1999) *The Internet in Everyday Life*, Cornwall: Blackwell Publishing.

Whitelegg, J. (1997) *Critical Mass: Transport, environment and society in the twenty-first century*, London: Pluto.

Wittel, A. (2001) 'Toward a Network Sociality', *Theory, Culture and Society*, 18: 51–75.

World Tourism Organization (2005) *Inbound Tourism: Tourist Arrivals by purpose of visit*. Online. Available at www.world-tourism.org (accessed 1 November 2005).

Zachary, G. P. (2000) *The Global M – New Cosmopolitans and the Competitive Edge: Picking Globalism's Winners and Losers*, London: NB-Publishers.

10 Getting into the flow

Peter Adey

> The noises of space. The colours of the world are coming toward me. I am plunged here and now in colors and noises to the point of dizziness. Here and now means that a flux of noises and colors is coming at me. I am a semiconductor. I admit it, I am the demon. I pull among the multiplicity of directions the direction that, from upstream, comes at me.
>
> (Michel Serres, *Genesis*)

Introduction

In this chapter I want to pick up on the title of the conference this collection is based upon: air 'time-space', for the reason that even while we see airports and airspaces as particularly fluid sites of transience, often the focus is directed towards the kinds of passenger mobility that flow, it seems, upon the surface of the airport's geography (see, for instance, Adey (2004) and the ground-breaking work of Crang (2002), Hannam *et al.* (2006), Fuller and Harley (2004), Gottdiener (2000), Cresswell (2006), Salter (2007; 2008, forthcoming)). Rendered simple containers through which people and things pass through, space and time are removed to a kind of passive stage upon which the drama of the terminal plays out, as opposed to an active and animate terrain that constitutes and is composed of practice, performance and social life (see also Adey (2006b) for this critique). One of the ways we might think about this is as a cinematic movie. It is as if airports have been imagined like the blue screens used to make films, where computer graphics and special effects are added in later. Just in this way, the airport is pushed to a backdrop – rendered a blue screen to be shortly replaced by other temporal and spatial contexts.

Readers might be aware of the poster advertisement for Steven Spielberg's fairly superficial but thought-provoking movie: *The Terminal*. The imagery of Tom Hank's character, Victor Navorski, is superimposed upon a white background that gestures towards Victor's in-limbo situation. If you have not seen the film, Victor's home country has undergone a civil war. This has rendered Navorski unable to enter the United States nor return to his own country. He is subsequently left – incapable of leaving the transit lounge of

New York's JFK Airport. And yet, the white background upon which Hanks is placed points towards something a little different. It relates to the way air time-spaces are dealt with in academic discussion. The whiteness resembles the absence of the geographies of such time-spaces. Hanks' character has been almost literally 'cut out' of the airport world because it doesn't matter. The poster illustrates a life on the move that has been literally shunted out of the times and places, the sensed and perceived geographies – messy environments of noise, dust, ambience, surfaces, odours, touch – in and with which, it will be argued, a mobile life happens (see also Wylie 2005).

It is easy to think of airports and airspaces through a contemporary time-frame. Indeed, I have argued elsewhere that popular approaches have tended to utilize a frame of what we might term a blind present (Adey 2006a), surely an easy thing to do as airports reflect the current political climate of security and fear just now. We must be careful not to understand airports as time-spaces of excess, existing and enveloping any time and any place with a one-dimensional mono-version of the airport. Removing or robbing airports of their history, as Massey puts it, works to immobilize the transience of these sites, stopping the airport dead in its tracks. To utilize Massey (2000: 228) again, it is as if the airports 'await our arrival. They lie there, in place, without trajectories.'

Of course, examining space and time is a well-rehearsed endeavour, and many recent writings have examined how time-space is continually called forth or brought into being, drawing upon the theorizations of *evolution, becoming* and *events*, as developed by theorists such as Bergson (1911), Whitehead (1979) and Deleuze (1993). Walking this line, geographer Mike Crang (2001) asks us to consider space and time as much more than simple terrains upon which the drama of life plays out, but as a timespace – lived and produced through action and practice. In this particular way, the following pages continue these sorts of effort by attending to just how social practices occur *through* and *with* this kind of animate space and time, as opposed to just on-top-of or without. Throughout the paper I argue that the movement of space should not remain merely the domain of philosophical arguments concerning the nature of matter, or indeed within facts of a microscopic physical world apparently irrelevant to human concerns (see, for instance, McCormack (2007) on the microscopic and Anderson (2007) on nanotechnologies). Rather, the case is made that these mobilities are experienced in a personal and social way. The movement time-space of airports inflects experience in a manner that is formative of identities, sensations and emotions (see also Thrift (2004) and Massumi (2002)).[1]

The paper is structured according to the several different ways that the flows which make up air time-spaces – particularly complex formations such as airports – are addressed and engaged with in embodied practice. The following section explores how flows are represented and talked about. Later sections investigate one's ability to coordinate and navigate within a space of flows, and then how one may preclude and make oneself apart from the flow. Finally, these issues throw up innumerable problems for conducting mobilities research. How, as researchers struggle with the same concerns, can we get into and make sense of the flow? The paper will draw the empirical examples together with the kinds of methodological challenge air time-spaces pose in the final part.

Talking flow

In many ways, airports are extremely immobile, sunk by the capital they embody in a way not dissimilar to other vast infrastructural systems, as Steve Graham (1998) – by way of David Harvey (1996) – tells us (see also Graham and Thrift (2007)). But, at the same time, airports not only see numerous flows passing through and by them, but they are – in themselves – part of the flow. Even within the theory of their internal design, airports are often intended to be rather fleeting. The first airport designs were premised upon fluidity, with many research papers in engineering and architectural journals celebrating themes of flexibility. The airport designs of the 1930s embodied the progress of modernity, as speed and plasticism materialized, not only in style and symbolism such as art deco architecture, but also in the workings of the terminal structure.[2] An article in *The Architects' Journal* of 1936 stated:

> in the present fluid state of air transport costly and permanent building schemes should be avoided. The whole lay-out should be one which will allow for ample expansion in case of need with the minimum of disturbance to the daily routine of the airport while additions are in progress. Ideally, the buildings should be on the principle of the sectional bookcase, capable of being enlarged by the addition of standardized units which will fit into the general scheme of the terminal building.

Airports of today rest upon a similar logic. Supplemented with the flows of food stuffs and consumables, passengers, visitors and staff, luggage, goods, fuel and more, the airport forms a nexus of movement (Graham 2003). For various authors commenting on these sites, which need no rehearsing here, the chaos of the terminal building makes it difficult for one to dwell within the constant movement and the hustle and bustle of people (Merriman 2004; Augé 1995; Castells 1996). But, let us consider that it is also the stuff – the *thingness* of the airport terminal and its movement – that problematizes one's capacity to dwell and make oneself at home, and how we make sense of these sites through, for instance, any attempt to represent them.

Take the issue of how something can be represented that is always on the move. Consider an attempt to represent an airport. Every time it is returned to, perhaps twice a week, perhaps next week, perhaps now, it has moved on; it is always different. What does it mean to say that one is researching an airport? Would this refer to the airport now, or the airport as it was? When writing about airports, would its representation refer to an airport recorded and captured in memory, or the version stored on the hard drive of a laptop computer? In other words, the incessant mobility of the site meant it is impossible to talk about it without a sense of uncertainty of what it was, and what it might become, following Alfred North Whitehead's famous maxim, 'There is no holding nature [and I might add the airport] still and looking at it' (Whitehead 1920: 15).

Of course, thinking about the airport and how we represent it in this way borrows greatly from the performative and non-representational styles of thought that have emerged in the wider arts and social sciences over the last decade, particularly in human geography (Dewsbury *et al.* 2002; Thrift and Dewsbury 2000). Taking cue from these works, my suggestion is not to say that the airport is unique in this way, but that airports are both useful reflectors of these ideas, while they also exacerbate them owing to the intensity of their movements relative to others.

But let us forget writing practices for now, for just talking about the airport offers similar difficulties. Asking airport staff to describe the airport and their role within it is met with confusion and frustration – related to their occupation and their attempt to describe it. Terminal Duty Managers (TDMs) tell me how they try to insert themselves into the flow. An interview with a TDM was continuously interrupted with the static and broken voices of the radio they were half listening to. They coordinate and synchronize their staff *with* passenger flows and needs. Staff rostering is carefully managed to coincide with busy periods; shop managers arrange their products, gifts and displays to match up with particular dates and times when particular passengers will be passing through. Circulation and its corollaries of uncertain encounters and events, as Foucault (2007) has shown, must be addressed by various technologies of anticipation. Not surprisingly then, the continual and uncertain emergence of these sites entails a level of complexity that must be solved by the calculative powers of software such as airport management operational support systems (AMOSSs), the sorts of system that Dodge and Kitchin (2004) tell us augment and premediate a space and a world that is always coming into existence (see also Grusin 2004; Thrift 2004; Dodge and Kitchin 2007).

Uncertainty also creeps into social relationships. For the duty managers to convey their role is a struggle of contextualizing their job within a continuously shifting spatiality. In conversations we try to place ourselves by fixing the airport to instances and moments of encounter. 'So was there scaffolding above the check-in area?' my respondent asks. I respond 'yes'. 'Ah so that must have been a couple of weeks ago', they respond. While this paints a rather positive picture, we often do struggle to find examples of when our versions of the building match up. In an interview with an airport architect, our 'emplaced' (Büscher 2006) knowledges of the building are recollected and compared with the rather different building plans in front of us, which have long since been altered and developed further. We recall when we were last there, moments when it was being extended, when building works were taking place. And this leads on to my next theme.

Trapping flows

Given all of this, if aeromobile spaces are so marked by this fluidity, mobility and change, why do we not sink through the floor, why can we not pass through the walls as easily as we might a wall of water? Indeed, if the airport system is marked by the control of movement, by containing mobilities, as Fuller and Harley (2004) tell us so well, how can this be if the airport's material environment is in continual flux? Let us consider and unpack, in the rest of the section, how one may navigate and negotiate this flow as a mobile body inhabiting a space of an ever-shifting character.

Following Sandford Kwinter's work on the relationship between architecture and temporality, I think the key is for the passenger, as in Kwinter's figure of the rock climber, to try to tap or trap the flows of the building, to 'insert himself [or herself] into a seamless, streaming space, and to subsist in it' (Kwinter 2001: 31). Moving, subsisting, dwelling, these seem to require a certain kind of mooring or anchoring.

But, understood in Kwinter's terms this needn't mean immobility. For Kwinter, climbing a rock face is an event that encompasses the meeting of two flows – the rock and the climber. Let me quote from Kwinter in more depth as he describes the relative fluidity of the mountain face. He explains:

> it is the mountainface itself whose flow is the most complex ... The mineral shelf represents a flow whose timescale is nearly unfathomable from the scale of duration represented by the electrolytic and metabolic processes of muscle and nerves – but even at this timescale – nanometric in relation to the millennia that measure geological flows – singularities abound: a three millimetre-wide fissure just wide enough to allow the placement of one segment of one finger, and anchored by sufficiently solid earth to permit but eighty pounds of pressure for, say, three seconds but no longer; an infinitesimally graded basin of sedimentary rock whose erratically ribbed surface (weathered unevenly by flows of wind and rain) offers enough friction to a spread palm to allow strategic placement of the other palm on an igneous ledge a half meter above.
>
> (Kwinter 2001: 31)

Thus, the mountain face is in flow and affected by flows just as the rock climber is. And yet, the rock face, until recently considered slick and featureless, is now changed. Phenomenologically it is different not because the climber necessarily changes it, but because of the climber's shifting posture towards the rock. Through the haptic engagement between the rock climber and the rock, the climber is able to find himself points and holds.

We might understand these permanencies and footholds, on the rock face and in the airport, as sorts of singularity: nodal points made by an assemblage of relations and forces. The passenger gives the terminal mobilities traction. It is in relation to the passenger that the terminal walls, floors and architecture are given their fixity and permanence, attaining form from within and without. Bodily capacities restrict one from knocking down the structure. Just enough force and a handrail acts as a guiding structure, too much and it becomes a broken eyesore. Human intuitions lead people unconsciously around the building as simple design functions anticipate bodily dispositions. A foot is placed down upon a limestone slab for an imperceptible moment, or a longer delay, before lifting off and moving on to the next. Such a relationship is often pre-engineered by the buildings' designers in order to encourage a certain route through the space. Holding these kinds of arrangements together, airport by-laws such as the code of 'displacement' constrain passengers from moving things. Getting into the flow, then, is not about an insufferable primacy given to fluidity and flux, but it is also about how relative and relational obduracies are made (Law 1994; Hommels 2005). The airport, in other words, is a product of force relations – the meeting of passenger and airport.

Navigation, for example, is contingent and dependent upon these sorts of relation. It is about existing and synchronizing with the air time-space. Signage systems change and are moved by informational flows – an informational architecture (Fuller 2002) – that updates flight information displays. They move because of grumpy customers insisting on alterations to the signage design. Or changes to the airport's design can scramble the flow chart-like stages of passenger processing, necessitating amendments to the signs. But for successful navigation, these signs must, according to one correspondent, be 'in the right place, and at the right time'; they must flow alongside the passenger in order to achieve the position of a relative stationarity. If we accept that an airport is a space of multiple itineraries and multiple temporalities, of 'multiple *durées* coming together . . . a place not necessarily of singular time but a particular constellation of temporalities' (Crang 2001 190), we see masses of people on their own timetables of scheduled departure or arrival times and connecting flights. Time is built into the airport terminal. This is a general and universal time. Passengers must be at the check-in hall x hours before their flight. They will be called through security x hours before departure. The gate is meant for passengers x minutes before boarding. Relative to the passenger, the sign must correspond with where the passenger is in their journey through the airport, and in the virtualized itinerary space that schedules *when* they need to be.

Stepping in and out

Now throughout this chapter so far dimensions of 'inside' and 'outside', 'within' and 'without' have been mentioned. This is for a good reason because how we are placed in relation to flow gives a sense of this insideness and outsideness, making a difference to experience and, furthermore, to our experience of the world's movement. In this section the paper considers how the atmospheres and substances of the airport are experienced quite differently according to how one is placed in relation to them.

Tim Ingold (2000) and geographer J. D. Dewsbury describe how we are subjects caught up in the world's becoming; human beings are aspects of the world's process. Ingold writes, 'in dwelling in the world we do not act *upon* it, or do things *to* it; rather we move along *with* it' (2000), just as J. D. Dewsbury (2000) describes the difficulty in apprehending a building's slow collapse as we dwell inside it. In other words, we are often already within the flow and, hence, may not realize it; we are subjects form*ing* together with the forms around us. Stuart Brand (1994: 4) deals with such a problem by asking us to, 'look up from this book, what you almost certainly see is the inside of a building. Glance out a window and the main thing you notice is the outside of other buildings. They look so static.' In short, our within-ness to flow can generate certain shapes and consistencies. Fixity and immobility might be understood as relational effects of our placing.

One may generate one's own awareness of the airport's changing material physicality by stepping outside of the usual temporal frames from which many passengers experience the site. As a researcher experiencing an airport for myself, I was not there to travel, nor was I following the usual 'time-path routines' of an airline passenger (as Hagerstrand might have put it). Shops were refurbished and rebuilt, seats and chairs replaced, floors resurfaced, shops restocked, advertising changed to cater for incoming flights. Following Lefebvre's rhythm analysis it seemed necessary to step out, to get outside the flow in order to apprehend it. He writes,

> When rhythms are lived, they cannot be analysed. For example, we do not grasp the relations between the rhythms whose association consti-tutes our body: the heart, respiration, the senses etc. We do not grasp even a single one of them separately, except when we are suffering. In order to analyse a rhythm, one must get outside it. Externality is necessary.
>
> (Lefebvre 2004: 88)

But stepping out is not always so easy. Naturally, one could get above, as Lefebvre tells us to – to see the space from the vantage point of elevated bridges and corridors. In this milieu patterns and rhythms emerge. The attraction to these sorts of pattern and this kind of activity is witnessed in practices such as people watching – a favoured activity among bored travellers (Adey 2007; Knitter 1996). And yet, some airport movements exceed a geographical location that may be simply left or climbed out of. Imagine what one might consider as airport noise, or the sounds that pervade the terminal. The sounds of plates clinking, footsteps, baggage wheels, talking, occasional shouting, doors opening. These are not localizable noises, for they move and disrupt other spaces and events, including the interview and the traces taken away in the form of an interview tape.

Furthermore, this issue is much more than a methodological quandary. For the hum, this background atmosphere can obviously be enjoyed as much as it can be disliked, and, for others, it may appear to be more of a foreground issue. As Serres writes, noise has the capacity to assail space, it resonates in us: 'we are utterly taken over by this murmer . . . our being is disturbed' (1995: 13). The noise, the chaos, can quickly overcome some people, erupting in states of panic, stress and sometimes anger. The evolution of an airport's 'bass materialism' (Goodman 2007), represented by ever-expanding noise contours, is a matter of discomfort for an airport's neighbours (see also Fuller and Harley (2004)). Such contours resemble the kind of amorphous, amoeba-like characteristics Reyner Banham once found so threatening (Pearman 2004; Banham 1962). Furthermore, the flows airports discharge, both atmospheric and chemical, pose numerous problems for environmental managers over long periods of time. Within the terminal itself, flows accumulate to create permanencies. Dust and dirt accrue in difficult-to-reach places to form persistently stubborn eyesores for observant passengers. In the next section I dwell on efforts to escape and get out of these oceans of interference.

Getting out

While we as researchers might be interested in getting out in order to observe, not better, but to get a different kind of experience, there are ways that aeromobile spaces are sites which people simply want to escape, where people decouple and disentangle themselves from the flow. Let us now consider the sorts of practice and design practice that enable this kind of separation in order to create senses of dwelling and homeliness.

If people search hard enough they will find that many airports contain their own quiet spaces; many airports have prayer rooms. But why might these be provided? Some studies have explained that it is possible for people to yank themselves away from the turmoil of places of travel. As Ursula Le Guin tells us, passengers might literally change planes of existence as opposed to the planes of the aircraft kind (2003). For geographer Julian Holloway, some people hope that shutting one's eyes or staying still may create a kind of 'space-time that can engender a spiritual connection' (Holloway 2003: 1968). Whilst this may be the case for some travellers, the provision and use of these spaces seem to suggest that, for others, this is simply not the case (for more on this see Kraftl and Adey (2008)).

Prayer rooms, multi-faith rooms and quiet spaces have been constructed in lots of airports in an attempt to provide for the yearning to get out of the flow, to leave the noise and the chaos of the terminal space. As airport architectural expert Brian Edwards writes, spaces are provided where 'a jaded passenger ferried from building to plane and terminal to gate, can find tranquillity and peace' (2005: 74). Indeed, many of these sites have been constructed with certain spatial characteristics that, I will argue, work to insulate and cushion the passenger from the flows discussed so far.

Far from harmonizing the space with the outside environment, many prayer rooms, gardens and quiet spaces have been designed to be distinctly apart and differentiated from their outside. Thus, prayer rooms are often tucked away in a small room apart from the main thoroughfare of passengers. Gardens may be placed in a small corner, sheltered by the side of the terminal building and out of the way of taxi and bus set-down areas. Business, club and first-class lounges are also found out of the way. Interestingly, Gatwick airport now features a Yotel within its terminal space. Designed as a pod-like hotel room/first-class airline cabin, the hotel user is cocooned away from the rest of the airport.

In discussion with a designer of one of these sorts of space, efforts to quell the intensity or the flux of the environment were given significant attention. What architectural theorist Peter Zumthor (2006) describes as atmospheres (see also Brennan 2003) might well apply to the moving materialities that can engender and communicate certain kinds of feeling, affects transmitted by the 'molecular energies' of sound, light and heat (see Conradson and Latham 2007). In this example, however, airports do not channel and circuit flows and energies along the kinds of line of Le Corbusier's biotechnical model (Fernandez-Galiano 2000); instead, they stop and halt them in specific places.

This is exemplified by one particular instance at the fast-developing Liverpool John Lennon Airport when the airport management was planning on adding a window to the prayer room. As an interior room, the prayer room is fairly dark and has no natural light. To an extent the isolation and insulation of the room had occurred by accident. Available space was at a premium, and it was one of the free areas of the terminal the airport could allocate. It was only when the airport management considered adding a window to the room that the Chaplain realized what a disruption it would cause to the room's feeling. On reading comments from the visitors' book – where visitors often signed and wrote down their thoughts and suggestions – the chaplain realised that adding a window would allow the movement and intensity of the outside world to rush in, spoiling the feeling of the room.

Passengers wrote that the room offered them a kind of 'oasis amongst the chaos' and rush of the terminal space. Comments centred on the 'relative peace' the space gave them away from the terminal. Visitors mentioned what the room felt like: 'warm' and 'cosy', 'quiet' and 'silent' featured highly. Thus, the chaplain halted the construction of the new window and attempted to manage the energies of the other flows that moved through the room; the lighting was muted, and a warm and atmospheric setting was created in stark contrast to other areas of the terminal.

For many passengers, spaces such as these offer the ability to step out. It is important then to realize that there are moments and spaces within transport interchanges such as airports or, as David Bissell (2007) shows, in train travel, of inactivity, stillness or suspension, when we need to remove ourselves from the flow and become isolated in relative fortresses of solitude.

Researching aeromobile spaces

In this discussion the airport appears as a perpetually moving tapestry of objects, people, things and information, interweaving each other and through one another. Practices, tactics or strategies to find one's way, to dwell within, or indeed to escape from such a place, involve a host of different ways of address and posture towards such flows. But of course, how one is placed in relation to the flux shapes the very way they are perceived and negotiated. Background shapes and flows rise to the surface in some instances. In others, the airport scrumples up and it folds; wrinkles of time-space may afford peaks of excitement, troughs of isolation or plateaus of boredom – all interrupting the apparently even flow.

Some of the characteristics described above offer particular issues for conducting research in these spaces. As discussed initially, the kinds of representational strategy we could use to write and talk about them must be questioned. Attempting to lock the airport down, to arrest it on pad and paper, text on page, and document and directory, may simply not suffice.[3]

This issue of how and what we take away is further problematized by that which we might not want. Going through my old interview tapes I came across an interview conducted in the back office of the information service desk. One of the most revealing issues is the noise. The conversation can hardly be made out because of the tannoy announcements – something airports are attempting to cut down on to reinvent themselves as 'silent airports'. As Michel Serres puts this fundamental problem, noise seems to settle 'in subjects as well as in objects, in hearing as well as in space, in the observers as well as in the observed, it moves through the means and the tools of observation, whether material or logical, hardware or software' (1995, 13). The pervasiveness of this noise poses the question: how can we draw out patterns from the chaos, the noise, the background radiation or interference? How can we separate anything from the constant hum of the airport machine? Or posed the other way, how can we get to the thrum, the noises and intangible atmospheres of the airport?

Moreover, how do we get to how these environments are experienced? Ethnography and participant observation might mean performing the practice of the airport journey – where one can experience the stresses and strains of being processed. To get into the flow could mean practising a form of self-reflecting flânerie performed by the writers Walter Benjamin adored. Like Baudelaire or Balzac, one might stroll the concourses of the airport, just as they – as *flâneur* – strolled the passageways of Paris. One looks out for the gesturing and gesticulations of fellow walkers. Indeed, for Benjamin (1973: 55), 'Empathy is the nature of the intoxication to which the *flâneur* abandons himself in the crowd.' In this way, stepping into the flow for the researcher can mean stepping into the shoes of those he/she walks with. One can be oneself, 'and anyone else as he sees fit'; for Marx one could fulfil the role of a ghost lusting for corporeal possession, 'Like a roving soul in search of a body, he enters another person whenever he wishes' (Marx in Benjamin, 1973: 55).

Conclusion

We can conclude that the multiple inhabitations and modes of passing through airports mark out sets of distinctive relationships that revolve around getting into the flow of things. In animating not only the foreground, but the airport environment itself, we can witness how bodies, objects and things move along with each other in different ways and at different times.

It should be noticed that these flows are not constitutive of some kind of simplistic notion of resistance or consumption. Indeed, this animated space is something recognized by the airports themselves, particularly in the way they are designed and managed. Luis Fernández-Galiano's (2000) summing up of the contradictory fluidities of architecture and energy nicely characterizes this sort of dynamic. He writes that architecture, 'can be understood as a *material* organisation that regulates and brings order to *energy* flows; and, simultaneously and inseparably, as an *energetic* organisation that stabilises and maintains *material* forms' (2000: 5).

Coming back to their inhabitation, processes of moving along with, trapping, getting into or indeed leaving airport spaces are suggestive of wider contemporary experiences in other mobile and animate worlds. But there is also something unique about them too: in the way these travelling companions relate and register with each other; in the way airports and aeromobilities have been taken within; in the way they are expressed in the drama, fears and emotions of the terminal and in the efforts being made to manage and capture these flows.

Finally, getting into the flow is also an issue for those conducting what is becoming known as mobilities research, as I have shown. Subsisting in, visiting, trapping, representing and talking about air time-spaces and other places are affairs that pose some new problems, questions and promising avenues for further research inquiry.

Notes

1   In so doing, I am following the work of thinkers such as Brian Massumi who, of course, draws upon Deleuze, Bergson and Whitehead.
2   Although the flexibility was not the only way progress was embodied. Art deco styles, in particular, came to symbolize an age of speed, movement and power (see Adey 2006; Pascoe 2001; Pearman 2005).
3   Of course such a problem remains tied to the, perhaps, almost impossible aim of attempting to represent something perfectly.

Bibliography

Adey, P. (2004) 'Surveillance at the airport: surveilling mobility/mobilising surveillance', *Environment and Planning A*, 36: 1365–80.
—— (2006a) 'Airports and Air-mindedness: spacing, timing and using Liverpool Airport 1929–39', *Social and Cultural Geography*, 7: 343–63.
—— (2006b) 'If mobility is everything then it is nothing: towards a relational politics of (im)mobilities', *Mobilities*, 1: 75–94.
—— (2007) ' "May I have your attention": Airport geographies of spectatorship, position and (im)mobility', *Environment and Planning D Society & Space*, 3: 515–36.

—— Budd, L. and Hubbard, P. (2007) 'Flying Lessons: exploring the social and cultural geographies of global air travel', *Progress in Human Geography*, 31: 773–91.

Anderson, B. (2007) 'Hope for nanotechnology: anticipatory knowledge and the governance of affect', *Area* 39: 156–65.

Augé, M. (1995) *Non-Places: introduction to an anthropology of supermodernity*, London; New York: Verso.

Banham, R. (1962) 'The Obsolescent Airport', *Architectural Review*, 250–60.

Bednarek, J. R. D. (2001) *America's Airports: airfield development, 1918–1947*, College Station: Texas A&M University Press.

Benjamin, W. (1973) *Charles Baudelaire: a lyric poet in the era of high capitalism*, London: NLB.

Bergson, H. (1911) *Creative Evolution*, New York: H. Holt and Company.

Berman, M. (1983) *All that Is Solid Melts into Air: the experience of modernity*, London: Verso.

Bingham, N. (2004) 'Arrivals and Departures: civil airport architecture in Britain during the interwar period', in J. Holder and S. Parissien (eds.) *The Architecture of British Transport in the Twentieth Century*, New Haven, Connecticut and London: Yale University Press.

Bissell, D. (2007) 'Animating Suspension: Waiting for Mobilities', *Mobilities*, 2: 277–98.

Brand, S. (1994) *How Buildings Learn: what happens after they're built*, New York, NY; London: Viking.

Brennan, T. (2003) *The Transmission of Affect*, Ithaca, New York, London: Cornell University Press.

Büscher, M. (2006) 'Vision in motion', *Environment and Planning A*, 38: 281–99.

Castells, M. (1996) *The Rise of the Network Society*, Oxford: Blackwell.

Conradson, D. and Latham, A. (2007) 'The Affective Possibilities of London: Antipodean Transnationals and the Overseas Experience', *Mobilities*, 2: 231–54.

Crang, M. (2001) 'Rhythms of the City', in J. May and N. J. Thrift (eds.) *Timespace: Geographies of Temporality*, London: Routledge.

—— (2002) 'Between Places: producing hubs, flows, and networks', *Environment and Planning A*, 34: 569–74.

Cresswell, T. (2006) *On the Move: the politics of mobility in the modern west*, London: Routledge.

Deleuze, G. (1993) *The Fold: Leibniz and the baroque*, Minneapolis: University of Minnesota Press.

Dewsbury, J. D. (2000) 'Performativity and the event: enacting a philosophy of difference', *Environment and Planning D-Society & Space*, 18: 473–96.

Dewsbury, J. D., Harrison, P., Rose, M. and Wylie, J. (2002) 'Enacting geographies – Introduction', *Geoforum*, 33: 437–40.

Dodge, M. and Kitchin, R. (2004) 'Flying Through Code/Space: the real virtuality of air travel', *Environment and Planning A*, 36: 195–211.

—— and —— (2007) 'Outlines of a world coming in existence: Pervasive computing and the ethics of forgetting', *Environment and Planning B*, 34: 431–45.

Edwards, B. (2005) *The Modern Airport Terminal: new approaches to airport architecture*, New York: Spon Press.

Fernandez-Galiano, L. (2000) *Fire and Memory: on architecture and energy*, Cambridge, Massachusetts; London: MIT Press.

Foucault, M. (2007) *Security, territory, population: lectures at the College de France, 1977–78*, Basingstoke: Palgrave Macmillan.

Fuller, G. (2002) 'The Arrow-Directional Semiotics: Wayfinding in Transit', *Social Semiotics*, 12: 231–44.

—— and Harley, R. (2004) *Aviopolis: a book about airports*, London: Blackdog.

Giddens, A. (1990) *The consequences of modernity*, Stanford, California: Stanford University Press.

Goodman, S. (2007) 'Bass Materialism', paper presented at Theorizing Affect, Durham.

Gottdiener, M. (2000) *Life in the Air: surviving the new culture of air travel*, Lanham, Md: Rowman & Littlefield.

Graham, A. (2003) *Managing Airports: an international perspective*, Oxford: Butterworth-Heinemann.

Graham, S. (1998) 'The end of geography or the explosion of place? Conceptualizing space, place and information technology', *Progress in Human Geography,* 22: 165–85.

—— and Thrift, N. (2007) 'Out of order – Understanding repair and maintenance', *Theory Culture & Society*, 24: 1.

Grusin, R. (2004) 'Premediation', *Criticism*, 46: 17–40.

Hannam, K., Sheller, M. and Urry, J. (2006) 'Editorial: Mobilities, Immobilities and Moorings', *Mobilities*, 1: 1–22.

Harvey, D. (1996) *Justice, Nature and the Geography of Difference*, Cambridge, Massachusetts: Blackwell Publishers.

Holloway, J. (2003) 'Make-believe: spiritual practice, embodiment, and sacred space', *Environment and Planning A*, 35: 1961–74.

Hommels, A. (2005) *Unbuilding Cities*, Cambridge, Massachusetts: MIT Press.

Ingold, T. (2000) *The Perception of the Environment: essays on livelihood, dwelling and skill*, London: Routledge.

Knitter, H. (1996) *Holding Pattern: airport waiting made easy*, Okemos, Michigan: Kordene Publications.

Kraftl, P. and Adey, P. (2008) 'Architecture/affect/dwelling', *Annals of the Association of American Geographers*, 98: 213–31.

Kwinter, S. (2001) *Architectures of Time: toward a theory of the event in modernist culture*, Cambridge, Massachusetts: MIT Press.

Law, J. (1994) *Organizing Modernity*, Oxford: Blackwell.

Lefebvre, H. (2004) *Rhythmanalysis: space, time and everyday life*, London, New York: Continuum.

Le Guin, U. K. (2003) *Changing Planes*, London: Harcourt.

McCormack, D. (2007) 'Molecular Affects in Human Geographies'. *Environment and Planning A*, 39: 359–77.

Massey, D. (2000) 'Travelling Thoughts', in P. Gilroy, L. Grossberg and A. McRobbie (eds.) *Without Guarantees: in honour of Stuart Hall*, London: Verso.

Massumi, B. (2002) *Parables for the Virtual: movement, affect, sensation*, Durham, North Carolina: Duke University Press.

Merriman, P. (2004) 'Driving Places: Marc Augé, Non-places, and the Geographies of England's M1 Motorway', *Theory Culture and Society*, 21: 145–68.

Pascoe, D. (2001) *Airspaces*, London: Reaktion.

Pearman, H. (2004) *Airports: A Century of Architecture*, London: Laurence King.

Salter, M. B. (2007) 'Governmentalities of an airport: Heterotopia and Confession', *International Political Sociology*, 1.

—— (2008) *Politics at the Airport*, Minneapolis: University of Minnesotta Press.

Serres, M. (1995) *Genesis*, An Arbor: University of Michigan Press.

Sudjic, D. (1992) *The 100 mile city*, London: A. Deutsch.

Thrift, N. (2004) 'Movement-space: the changing domain of thinking resulting from the development of new kinds of spatial awareness', *Economy and Society*, 33: 582–604.

Thrift, N. and Dewsbury, J. D. (2000) 'Dead Geographies – and how to make them live', *Environment and Planning D-Society & Space*, 18: 411–32.

Whitehead, A. N. (1920) *The Concept of Nature. Tarner lectures delivered in Trinity College, November 1919*, University Press: Cambridge.

—— (1979) *Process and Reality: an essay in cosmology*, New York: Free Press; London: Collier Macmillan.

Wylie, J. (2005) 'A single Day's Walking: narrating self and landscape on the South West Coast Path', *Transactions Institute of British Geographers*, 30: 234–47.

Zumthor, P. (2006) *Atmospheres*, Basel: Birkhauser.

Part IV

Governing air travel

11 Science, expertise and local knowledge in airport conflicts

Towards a cosmopolitical approach

Guillaume Faburel and Lisa Levy

The transformation of societies into transnational social formations goes along with fundamental and inevitable conflicts. 'Global complexities' and social structures 'beyond societies' (Urry 2000; 2003) often become visible as specific risk management policies – not at least in the transnational transport infrastructure system. The relations between airports and the localities that host them are basically shaped by conflicts that are characteristic of the mobilization and cosmopolitanization of modern societies (see Kesselring (Chapter 2), in this volume). They also form the spatial shape of airport regions. Hence, the 'geography of noise' (Faburel 2001) of airports signifies the territorial impact of globalization and increasing mobilities. Among many other hotly disputed issues, noise attracts great attention, despite various regulatory measures implemented over the past thirty years, such as acoustic aircraft certification, noise thresholds in the vicinity of airports, and sound-proofing subsidies. As demand for air travel continues to grow, resulting in the expansion of air traffic and airports, we are witnessing a proliferation of groups, movements and campaigns protesting against such expansion and its local environmental impact.

Because these conflicts increasingly impinge on the future of air travel, decision-makers begin to attend to the issue of the social acceptability of airports to the local communities where they are located. And since aeromobility is growing more rapidly than the technologies that seek to minimize its negative impacts, there is a mounting need for complementary and/or alternative measures, notably environmental ones, and particularly regarding major international airports. As a result, environmental impact issues are beginning to structure the future of aeromobility, with nuisance caused by aircraft noise playing a major role in this process.

Noise reduction or preventative measures are already constraining factors in several European airports (Eurocontrol 2001). Nevertheless, the idea of internalizing the social costs of aircraft noise is again resurfacing as a major focus in the debate, alongside the question of aircraft noise impact assessment. The environmental externalities of air transport have rarely been internalized. Admittedly, there are several dozen international airports that tax take-off noise, adjust landing fees, or even impose fines for non-compliance with nominal schedules and trajectories (Morrell and Lu 2000). However, although these measures are all linked to the polluter pays principle (PPP), no airport gauges the amount of this tax or aligns fees to the true impact of noise.

Moreover, actions invoking the PPP, although consistent with the internalization of the social and environmental costs of aviation, have led to a proliferation of scientific models and methods of evaluating these costs. This has resulted to a great extent in evaluations of the social costs of aircraft noise or wider environmental air travel impacts generally being neither implemented nor followed up at the institutional level. There is a crucial lack of consensus about the evaluation of the social and environmental costs of aviation and particularly about the role that science and expertise play in the internalization of these costs as well as in the emergence of airport-related conflicts. This follows the fact that scientific knowledge and evaluation procedures are now concerns not only for those in charge of their production (the experts) but also for their objects (those who live in the vicinity of airports). In this way this provides an example of the new 'global complexity' (Urry 2003) of the 'mobile risk society' as Kesselring (2008; Chapter 2, in this volume) puts it.

The overall aim of this chapter is to position official knowledge systems in the context of the politics of air travel, looking at what characterizes decision-making in new governance structures that evolve around airports and air travel. More particularly, this chapter examines three interrelated issues: (a) the diminishing capacity of central authorities to generate and regulate collective action, which becomes increasingly fragemented and subjected to continuous negotiation (Gaudin 1999); (b) the emergence of new notions and categories that begin to characterize discourses about, and the politics of, air travel, such as sustainable development, governance, participatory democracy, environmental justice and territorialization; and (c) the changing socio-political role of scientific expertise (Duran 1999), in particular technical-economic evaluations implemented as parts of transport projects.

The chapter is based on two distinct but linked pieces of research. The first is a study conducted for the French Ministry of Ecology, Development and Sustainable Planning (Faburel and Mikiki 2004), which included interviews with various stakeholders at the Roissy Charles De Gaulle airport, Paris. The second is a more recent, wider in scope and multidisciplinary research project, involving nine international airports and conducted for the Centre National de la Recherche Scientifique and Aéroports de Paris, which operates several airports and airfields in and around the French capital. This research was based on interviews with around 150 stakeholders and investigated participatory procedures and debating forums that have emerged in several major airports, looking in particular at the dynamics of expert opinion and the implementation of knowledge systems, the emergence of sustainable development as a new vector shaping relations between stakeholders, and emerging spaces of collective action (Faburel *et al.* 2007).

Both these projects have revealed the shifting nature of the socio-political context of scientific knowledge about, and technical evaluations of, the social and environmental impact of aviation. In particular, they reveal how the instability of such knowledge led major players in the air transport industry to seek dialogue with stakeholders in the local communities that host those major airports, among them local authorities, grass-root movements and associations representing the local economy. This emerging dialogue in turn highlights the limitations of the traditional model of air transport planning and traffic management (based on *command and control*), leading to a shift in perceptions about the environmental impact of air travel as well as a trend towards more participatory decision-making processes concerning aircraft noise management.

The next section of the chapter examines traditional forms of knowledge about the social costs of aircraft noise, with a focus on techno-scientific discourses. The third section examines the emergence of alternative, territorial forms of knowledge, looking at their struggle against hegemonic discourses and for the legitimation of local contexts in the evaluation of the environmental impacts of airports. Then, by way of conclusion, we consider the emergence of a new system of expertise and governance within airport conflicts, as well as a novel, cosmopolitical approach to airport noise.

Science, technology and the impact of aircraft noise

One of the reasons long invoked by authorities for not internalizing aircraft noise impact was that social-costs assessment contained many indeterminate and unknown factors (Comité des Applications de l'Académie des Sciences 1999). Yet attempts to evaluate the social costs of transport noise have multiplied over the past twenty-five years (Navrud 2002). At the same time, assessment procedures have grown more sensitive, statistical modelling has become more elaborate, and results appear to be more reliable than before, prompting many to advocate their safe use in evaluations of airport construction or extension projects (Button 2003).

At Amsterdam's Schiphol airport, in particular, many studies have estimated the social cost of aircraft noise, one of which included a rating of the local population's well-being (Van Praag and Baarsma 2000). Another took an empirical approach to property depreciation imputable to aircraft noise (Morrell and Lu 2000), combining devaluation assessment with meta-analysis (Schipper 1997; Schipper *et al.* 2001). Although these initiatives did not lead to concrete solutions, they have revealed the disparity that exists between noise taxes and real social costs (Lu and Morrell 2001). Some of these studies have even suggested compensation to local communities (Baarsma 2001). In the case of London Heathrow airport (Pearce and Pearce 2000), the gap between social costs and real taxes led to the British government's decision to establish a threshold of aircraft movements at the new Terminal 5. On the whole, the negative impact of aircraft noise (variably characterized and measured in terms of nuisance, depreciation of property values and health problems) has invariably thrust the issue of internalization of social costs into the politics of air travel. And the perceived disparity between such impact and the related compensatory measures remains one of the keystones of protests against local aircraft noise.

Discussions about the impact of aircraft noise are not restricted to measures that implement the PPP. Other airports, such as Minneapolis St-Paul, San Francisco International, Vienna International, Los Angeles International and Sydney Kingsford Smith, have addressed the impact of noise on local communities through a number of different initiatives, such as the elaboration of alternative indicators in addition to those measuring emission or exposure, the extension of territorial perimeters for sound-proofing subsidies, the creation and expansion of airport-community noise committees, and the funding of alternative local expert counter-opinions.

Therefore, there has been great diversity in methods of evaluating the impact of noise around airports, and no less diversity in measures of internalizing the perceived social costs of such impact. Despite this apparent variety, there have been very few studies of the relationships between airports and their local communities (see Vallet *et al.* 2000; Faburel 2001; 2003a; 2005; Faburel and Barraqué 2002; Martinez 2001; and Periañez (2001) for the French context), and only recently have we began to understand both the dynamics of knowledge production and the evolving political struggles that develop in this field. Our research has sought to investigate in more detail how different discourses about aviation noise impact are formed, and how they get incorporated into the political arena by various stakeholders. We have thus identified different models of production and implementation of knowledge concerning the impact of aviation noise.

The model that has enjoyed predominance over the past two decades could be termed *technological legitimation*. This model is based on techno-scientific discourses, focusing on the validity of research and expertise, and tends to reject a number of aircraft noise impacts on principle. With side effects that are 'not proven' and studies that are not 'convincing', this model considers social cost evaluations of minor interest at best, highlighting the uncertainties that afflict such assessments. Even *nuisance*, the only effect of aircraft noise that has recently been universally accepted as harmful to individual health, is swiftly put into perspective as 'complex', 'subjective', 'emotional' or 'irrational'.

This model, supported chiefly by stakeholders in the air travel industry and central authorities, relies on objectivization through technology, backed up by a particular representation of the validity of knowledge that singularizes disciplines that are 'pertinent' to the subject (e.g. psycho-acoustics) as well as relevant scientific methods (e.g. multivariant statistics, economometric analysis, modelling techniques). This objectivization is politically productive: in the face of local 'passionate reactions', it rationalizes the debate by legitimizing certain arguments and modes of action and delegitimizing others. For example, these stakeholders believe that measurement outcomes for energy and exposure to sound are much more 'practical' and 'acceptable' than internalizing taxes that could affect the evolution of the air transport industry differently.

With a strong emphasis on the physical sciences and technology, this model naturally encourages actions based on acoustics, overlooking cultures, codes and regulations. The result is ever greater levels of technical sophistication in the representation of noise generated by airports, from Internet broadcasts of flight trajectories and acoustic imprints of every aircraft in real time (San Francisco, Frankfurt), to increasingly sophisticated maps (Sydney and Geneva) and larger numbers of measurement stations (Chicago O'Hare and Denver International). Technical prowess and aesthetically high-powered communication tools reign supreme. We see this trend notably in the United States (Faburel 2003b).

This technological model (through its chief scientific domain, acoustics) also establishes elements for particular regulatory frameworks espoused by the aviation industry, providing an international certification bulwark against any commercial discrimination of airlines, standard norms and procedures as used by overseeing regulators (such as the Federal Aviation Administration in the USA) as well as aids for the standardization of indicators in acoustic monitoring and other technical services of environmental regulators. In the same logic of this technology-based legitimation (acoustic or other), we witness the proliferation of certifications, benchmark indicators, environmental performance reports and other mechanisms of standardization and comparison of acoustic impact, all perfectly attuned to the technical-normative logic supported by the aviation industry. This in turn consolidates the industry position within political-administrative systems, one effect of which is the lack of research into the socio-spatial impact of air transport.

Despite the focus on the physical sciences, this model also influences the way that social, economic and political debates are framed. In particular, we have examined the political codes that have long determined the (non) evaluation of social costs and, at a deeper level, of aircraft noise impact, codes that reveal the structuring effects of the technical and functional representation of aviation phenomena, especially of their social and environmental impacts. For instance, in light of widespread recommendations by a number of influential economists for the internalization of external costs, monetary valuation of noise is still not used for determining tax, fees or rates, because of the uncertainties regarding monetarization, as set against the 'precision' of acoustics.

More generally, however, these political codes are related to the epistemological-republican dogma highlighted by Latour, which elevates certain scientific disciplines to the rank of normative values. This leads to the uniformity of scientific validity and dangerously reduces the value of scientific controversy. To this day, the use made of acoustics, psycho-acoustics and marginalist economics by official expertise and in the discourses of airports is part and parcel of such judgements, the effect of which is to mutilate diversity and to legitimize one single epistemology (technical) and one single political objective (reasoning with emotions).

This political vision of 'good' science presents itself as scientific. Its corollary is to divide up reality, with scientific facts on the one side, and socio-political values on the other. The separation maintained between these two airtight compartments is a necessary condition for the 'objectivity' of both data and measures. Conversely, this political vision of science considers socio-political compromise as a form of surrender: 'In this scheme of things, science is part of the solution of the political problem, the solution of which it renders impossible by continuously threatening to disqualify human assemblies' (Latour 1999: 59). Supported by this *technological legitimation,* traditional airport expertise has disqualified local stakeholders by delegitimizing the study of local impacts of aviation noise, making it harder for these to become subjects of both observation and debate.

Local communities and spatial representations of aircraft noise

Developing quite outside the techno-scientific model analyzed above, one finds a myriad of representations of the local impact of aviation noise. Although based on considerably different approaches, these representations point to the emergence of a new model of knowledge production and action in the field of airport noise. Described with reference to a number of different notions and concepts, among them 'identity', 'perception', 'experience', 'community', 'cooperation', this model can be best conceptualized as *territorial.* It suggests shared values and interests of people who inhabit the *local* scale of airports and their surroundings.

This model calls for actions that take into account the collective needs of local communities and the sustainable development of airports. Its approach to internalization calls for extra funds to be invested in those communities and for more conciliatory measures, balancing noise nuisance by way of repara-tion and compensation. Among such measures and policies are the allocation of airport jobs for local residents, subsidies for training, and creation of observatories of real-estate values.

With this model we witness the emergence of the territorialization of both knowledge and action. Local communities and associations make use of experiences generated by their respective itineraries and established social and political networks in order to build alternative representations of their places and introduce them into the debate. At the heart of this process lies a knowledge of planning, development, environment, local road traffic, public transport, etc., that local communities claim to be neither necessarily nor readily given to central governments, aviation authorities and airport operators. These claims lead to a justification of these communities' right to control and monitor, a right they feel deprived of. They also produce knowledge that is gradually obtaining recognition from other stakeholders.

Los Angeles International airport provides an interesting example of this process, where a powerful coalition was formed, uniting different local com-munities and associations in the El Segundo area (Coalition for a Truly Regional Airport Plan). With the help of urban studies specialists, this coalition sought to develop its own expertise in order to respond to arguments put forward by the air transport industry in the course of impact studies. The role given to local counter-opinions is partly linked to the Community Agreement reached in 2006 between this coalition and the City of Los Angeles, owner of the airport. Here and in other airports around the world, such communities sponsor studies of property value depreciation and health effects of aircraft noise, among other subjects, highlighting issues of identity and belonging, thus producing a kind of knowledge that is lay, local and practical.

While the techno-scientific model generally overrides particular spatial contexts, cultural specificities and modes of governance, the territorial model demands impact indicators that are more in line with the lived experiences of local residents, questioning the emphasis on acoustic metrics and the relative absence of social costs from representations of the environmental impact of aviation. This evolving politics of knowledge provides a key element for understanding the emergence of airport conflicts during the past few decades.

In opposition to the hegemonic model, local stakeholders suggest new ideas and analytical categories (e.g. social equity, environmental justice) as well as alternative research methods (e.g. surveys) that are more likely to represent the ordinary, local experiences of aircraft noise. During a mediation process at Frankfurt International airport, representatives of local associations vented their frustrations with existing expertise about the airport's local impacts: '[the experts] come here, they look around and deliver their results three days later . . . while declaring that their work is scientific.' In contrast, new evaluation instruments can provide bases for dialogue, especially if local residents participate in the processes of evaluation and monitoring (see Kesselring (Chapter 2), in this volume).

One particular knowledge strategy of local groups consists in providing 'anti-expert opinions' through detailed analyses of the conditions under which the expertise sponsored by the champions of airport expansion is produced. In Amsterdam, for instance, environmental and residents' associations from several international airport areas in Europe, especially London and Amsterdam, got together to commission such a counter-opinion from a research and policy outfit in Delft. In Frankfurt, the Rhein-Main institute, with the support of the Zukunft Rhein-Main association, brought together a group of eight researchers in September 2006 in order to evaluate the methodology used for a study that central authorities have used to justify the construction of a new runway. However, the critique of official expertise and the emergence of counter-opinions are not devoid of contradictions and difficulties. Many local stakeholders, especially in local authorities, hesitate between fighting aircraft noise and capitalizing on the tangible tax income that airports generate.

The constitution of the territorial model of evaluating the environmental impact of aviation provides expertise with an alternative source of legitimacy, namely new forms of local knowledge and know-how. It produces new spatial representations and introduces new experts into the field of airports and air transport, especially from the human and social sciences. These new forms of knowledge allow for alternative modes of expression and representation by local stakeholders, beyond the limitations of protest and direct action.

However, the territorial model has encountered difficulties in establishing itself as a real alternative. Historically, because it has had to react to the logic of technical objectivization and reification of effects that are at the same time social, and to the uniformity of regulatory norms over different territorial contexts, the consequence is that the hegemonic techno-scientific discourse remains the major reference in debates and conflicts over airport expansion and construction. The reactive nature of the territorial model is illustrated by the reception of debates sponsored by the Commission Consultative de l'Environnement at Roissy CDG (Leroux *et al.* 2002) and the Regional Dialog Forum in Frankfurt (Sack 2001; Geis 2003; Lévy 2005; Kesselring 2007). Residents' associations have described studies such as these, which aimed to nourish debate and defuse latent conflicts, as technocratic and devoid of deeper interest in territories and their experience.

One of the consequences of this is that, despite the alternative discourses of local stakeholders such as residents' groups, who call for greater territorial-ization of knowledge and modes of intervention, the impact of aircraft noise on airport surroundings remains largely under-investigated, while local iden-tities find it hard to establish themselves as legitimate issues. This lag derives in part from methods of pricing the local costs of aircraft noise. Myths about the structuring effects of mobility and transportation (Offner 1993) separate what should be kept together (namely, airports and their host communities) and perpetuate the strictly functional representation of airport infrastructure, equipment and surroundings, consolidated in notions such as 'hub', 'gateway' and 'hinterland'. This particular coding, which links systems of knowledge and value, has contributed to a collective blindness to the potentially complex and multiple effects of noise on local people.

One of the questions that needs to be asked is: As local spaces and place dynamics (e.g. residential mobility, spatial organization and social practices), and their multiple attributes (urbanistic, residential, political) finally enter the airport scene, why is it that sciences and knowledge forms that are best placed to represent this trend are slow to respond? And why is it that even today local communities rarely turn to them? We suggest two answers: First, terri-torial (local) stakeholders, even when producing their own counter-discourses, need to operate within legal constraints. This, and the scope of the airport environmental issues, requires that they deploy the evaluation capacities and resources of central authorities in the political-administrative system. And second, these stakeholders face charges of NIMBYism from both industry and central authorities, who seek to disqualify and undermine residents' claims and demands.

This charge creates a serious problem for these stakeholders, since it forces residents to seek to detach themselves from the experiences of their own localities in order to reach a higher level of generalization (Lafaye and

Thévenot 1993), which is necessary to legitimize their interests (Lolive and Tricot 2002). One strategy involves the extension of spatial or temporal scales. In the case of London Heathrow, for example, those who opposed the construction of a new terminal and runway made reference to the impacts generated by climate change caused by air traffic emissions. Everywhere references to future generations and their rights and to remote populations are being used as bases for argumentation, contributing to the removal of local referents from the debate and from the production of expertise.

Despite these limits, the uniformity of a normative position powered by the will to rationalize, and based on 'universal' acoustic knowledge, clearly helps generate conflicts whereby, in spite of different demands and modes of protest, action is invariably inspired by experiences of place. These experiences call for an opening of the dominant system of expertise to new forms of knowledge and new modes of governance that are more sensitive to the territories flown over. However, the logic of place (and its deployment of spatial proximity, living experience and concern for the common good) is also complemented by the constitution of inter-airport networks and umbrella organizations, following the reticular structure of air traffic distribution.

In the context of representations of the environmental impact of aviation, and in particular of noise around airports, the analytic sciences have contributed to a lack of anchoring of official knowledge in local contexts. However, and considering the importance of environmental issues for the future of aeromobility, the scientific practice of circumscribing the question of noise within technical referents appears to be, if not completely gone, at least relativized. The fact is that 'efforts in the past, that largely relied on engineering approaches to confront conflicts between social and environmental sustainability have failed, but nothing has yet emerged to fill the gap' (Button and Nijkamp 1997: 218). There is, however, a new pluralist expertise evolving around issues of aircraft noise and its effects on local communities, which will be the object of our concluding section.

Towards a cosmopolitical approach to airports' social and environmental impacts

Taking the issue of the social costs of aircraft noise as a starting point, it seems that monetary valuation can have real operational implications. Economic models and methods can provide more contextualized approaches that can help assuage local concerns through representations of airports' positive impacts on property values (in part due to increases in jobs and local transport services). However, the expectation of many local stakeholders runs high in airport conflicts, so room must be found for the evaluation of the social costs of aircraft noise. At this point, it would be useful to make greater use of knowledge generated by the human and social sciences, primarily sociology, social psychology, anthropology, social geography and political science. These disciplines stand at the very opposite end to acoustics, psycho-acoustics and

– to a lesser degree – marginalist economics (Barraqué 2003), and can rarely be found in this field, excepting perhaps the analysis of sound phenomena in the urban environment.

This mobilization and coupling of different knowledge traditions would make it possible to highlight and explain certain paradoxes that have become apparent in airport conflicts: while ground sound levels around some major airports have tended to remain more or less stable, or even – according to official data – decrease, nuisance seems to increase, as revealed by some longitudinal analyses (see Katska 1995). In particular, this coupling may help consolidate the monetarization of the social and environmental costs of aircraft noise, through, among other things, understanding the link between *ex post* satisfaction of households and the sound-proofing of homes in the case of the protection costs method, forecasting the health impacts of noise in the harm costs method, measuring annoyance in the contigent valuation method (Navrud 2002; WHO 2004), or observing criteria for residential choices in the hedonic prices method (Lake *et al.* 1998).

However, in this prospective emergence of objects of collective learning, apart from including other forms of knowledge and benefiting from their explanatory potential, it is necessary to facilitate interdisciplinary research in order to represent better the complexity and socio-spatial intertwining of aircraft noise impact. This effort would also benefit from scientific controversy, which would test preconceived and commonly used notions that have dominated the field. Such notions put a strain on the relations between stakeholders, more than they contribute to project common interests and compromises. For example, the aviation industry has long claimed that airport neighbourhoods attract new residents. It did so in the first place to oppose the results of fine-tuned observations of micro-spatial dynamics on the municipal scale. Then, when they finally admitted outward movement existed, they invoked the responsibility of local authorities and their 'lax' planning, and also the 'carelessness' of households in making their residential choices. Interdisciplinary approaches can, at least in part, modulate the still dominant model of linear, causal and deductive representations of the techno-scientific approach to these phenomena. They might also help overcome the technicist rut in which traditional expertise is still stuck to this very day (Faburel and Mikiki 2004), and contribute to the deconstruction of rather tenacious myths.

On the other hand, there is an inextricable link between the production of rationality and the exercise of democracy (Stengers 1997). New evaluation initiatives will not develop without new forms of political action and representation. It remains a fact that the question of noise has not been raised in numerous airport contexts. Although the acoustic basis of statutory instruments is the focus of the current debate, there are also political barriers to the expression of local noise experiences. There are great limitations to the technological framing of the debate: residents cannot be represented by decibels, nor can neighbourhoods and local authorities be represented by statutory zoning based on acoustics alone. There is the need to establish the *political pertinence* of expertise, not just its technical effectiveness. As environmental decisions increasingly turn to intermediary modes of regulation (e.g. contracts), the implementation of deliberative procedures (Callon *et al.* 2001) regarding the 'externalities' of airport operations might go in this direction as well. There are numerous indications of this: Vienna International and its control panel of sustainable development indicators, Los Angeles International with its Community Agreement, and, more recently, Orly's move to set up a sustainable development charter are just a few examples.

This democratic opening of procedures could open expertise to less recognized sciences or forms of knowledge, specifically local ones that are sensibility-oriented, practical and vernacular. These local forms of knowledge cannot be represented by the traditional sciences (even the social sciences). They correspond to a rationality and a mode of argumentation that never seek detachment. On the contrary, they define themselves via their attachment or proximity to their object. Their legitimacy is founded upon commitment, an ordinary sense of justice, a confrontation with local circumstances, and the identities of places subjected to aircraft noise.

Faced with the opening of borders (territorial, political, sectoral) engendered by mobility, notably aeromobility, their aim is to return to the scale of the surrounding territories, to the anchoring and identity of place. This return, far from being contradictory or regressive in terms of scales or stakes (political, economic, social), can be fully integrated into a cosmopolitical approach (Stengers 1997; Latour 1999) applied to expertise. When one avoids the compartmentalization of spaces and interests, the spaces of the airport and its surroundings appear as a source of fertile perspectives and actions. Born out of the reciprocal relationships between the humans and objects that constitute them, these spaces are by their very nature hybrid and interactive. Following a cosmopolitical approach, local struggles against airport noise point towards demands to slow down, to suspend movement, consequently making room for decisions based on forms of knowledge hitherto not sufficiently examined because of, among other things, lack of time (Lolive and Soubeyran 2007).

This slowing down may function as a counterpart to the acceleration produced by aeromobility. Far from giving rise to simplified or narrow perspectives, it would multiply the beings (human and non-human) concerned with this aeromobility. From a purely socio-political perspective, this opening could also – by taking into account these other forms of knowledge – contribute to new syntheses in the representations of various objects (e.g. noise impacts). However, research into airport conflicts reveals that such syntheses require the full recognition of the discourses conveyed by territorial stakeholders, endowing residents with true expert status, whose specific areas and forms of knowledge complete and complement the products of social science (Thévenot 2006).

The assertion of composite (or hybrid) expertise would contribute to transforming a 'bald', or 'riskless object' (aircraft noise), into a 'hairy object' or a 'risky attachment' (Latour 1999: 40). Approached from this angle, airport noise, in its essence, would in fact no longer 'have fixed and indisputable borders . . . with which no negotiation could succeed since the only thing one might expect from the proposals would be to tire the adversary' (Latour 1999: 129). Its expected and unexpected consequences can only be analyzed by a mobilization of the human and social sciences in association with the residents' practical knowledge. This object would then be defined by its multiple links with numerous beings (including residents, places and public authorities) and – confirmed as a co-constructed object – take part in the construction of a world in common where the complexity of airport noise and other environmental impacts of aviation, and aeromobility in general, are recognized.

Bibliography

Baarsma, B. (2001) 'Monetary Valuation of Noise Nuisance Around Airports: The Case of Schiphol', *Internoise Proceedings*, The Hague: 8.

Barraqué, B. (2003) 'Bruit des aéronefs: formule mathématique ou forum hybride?', *Espaces et Sociétés*, 115: 79–97.

Button, K. (2003) 'The potential of meta-analysis and value transfers as part of airport environmental appraisal', *Journal of Air Transport Management,* 9: 167–76.

—— and Nijkamp, P. (1997) 'Social Change and Sustainable Transport', *Journal of Transport Geography*, 5(3): 215–8.

Callon, M., Lascoumes, P. and Barthe, Y. (2001) *Agir dans un monde incertain. Essai sur la démocratie technique*, Paris: Le Seuil.

Comité des Applications de l'Académie des Sciences (1999) *Evaluer les effets des transports sur l'environnement, le cas des nuisances sonores*, Paris: Tec & Doc.

Duran, P. (1999) *Penser l'action publique*, Paris: L.G.D.J. Coll. Droit et Société.

Eurocontrol (2001) *Study on constraints to growth*, European Civil Aviation Conference, Vols. 1 and 2, Brussels.

Faburel, G., Rui, S., Déroubaix, J.-F. and Lévy, L. (2007) *Aéroports, environnement et territoires (AET): arènes de débats et modes de régulation des conflits. Retour d'expériences étrangères (Europe et Etats-Unis)*, Report of Centre for Research on Planning: Land Use, Transport, Environment and Local Governments (Université Paris XII) to French National Center for Scientific Research, and Aéroports de Paris.

Faburel, G. (2001) *Le bruit des avions. Evaluation du coût social*, Paris: Presses des Ponts et Chaussées.

—— (2003a) 'Le bruit des avions. Facteur de révélation et de construction de territoires', *L'Espace géographique*, 3: 205–23.

—— (2003b) *Les conflits aéroportuaires aux Etats-Unis. Lorsque l'approche technique de l'environnement conduit les aéroports dans des impasses*. Scientific stay at M.I.T. (Cambridge, US), Report of Centre for Research on Planning: Land Use, Transport, Environment and Local Governments (Université Paris XII) to French National Center for Scientific Research.

—— (2005) 'Quality of Life Near Airport. The Case of Annoyance and Property Values Depreciation near Orly Airport: Does the Aircraft Noise Exposure Really Lead to Understand Problems?', paper presented at *American Institute of Aeronautics and Astronautics*, Aircraft Noise and Emissions Reduction Symposium, 24–26 May 2005, Monterey California.

—— and Barraqué, B. (2002) *Les impacts territoriaux du bruit des avions. Le cas de l'urbanisation à proximité de Roissy CDG. Ne pas évaluer pour pouvoir tout dire, et son contraire*, Report of Centre for Research on Planning: Land Use, Transport, Environment and Local Governments (Université Paris XII) to French Environmental Agency.

—— and Mikiki, F. (2004) 'Valuation of Aircraft Noise Social Costs: Policy and Science Implications', paper presented at *World Congress of Transports*, July 2004, Istanbul.

Gaudin, J.-P. (1999) *Gouverner par contrat. L'action publique en question*, Paris: Presses de Sciences Po.

Geis, A. (2003) *Umstritten, aber Wirkungsvoll: Die Frankfurter Flughafen-Mediation*, HSFK-Report 13/2003.

Katska, J. (1995) *Longitudinal on aircraft noise. Effects at Dusseldorf airport, 1981–1993,* ICA: 15, Proceeding Trandheim.

Kesselring, S. (2007) 'Globaler Verkehr – Flugverkehr', in O. Schöller, W. Canzler and A. Knie (eds.) *Handbuch Verkehrspolitik*, Wiesbaden: VS Verlag.

—— (2008) 'The mobile risk society. Mobility and ambivalence in the second modernity', in W. Canzler, V. Kaufmann and S. Kesselring (eds.) *Tracing Mobilities*, Burlington: Ashgate.

Lafaye, C. and Thévenot, L. (1993) 'Une justification écologique? (Conflits dans l'aménagement de la nature)', *Revue française de sociologie*, 34(4): 495–524.

Lake, I., Lovett, A. A., Bateman, I. J. and Langford, I. H. (1998) 'Modelling environmental influences on property prices in an urban environment', *Computers, Environment and Urban Systems*, 22(2):121–36.

Latour, B. (1999) *Politiques de la nature. Comment faire entrer les sciences en démocratie*, Paris: La Découverte.

Leroux, M. and Amphoux P. (coll. Bardyn J.-L.) (2002) *Vers une charte intersonique. Préfiguration d'un outil interactif de diagnostic et de gestion des représentations de la gêne dans un système d'acteurs*, Report CRESSON to French Environmental Agency.

Lévy, L. (2005) *Conflits et planification autour des grands aéroports internationaux: vers un développement durable des territories aéroportuaires? Etude comparée des cas de Francfort Rhin-Main et Roissy CDG.* Mémoire de maîtrise sous la direction de G. Faburel, 184 pp. (Univeristé Paris 1).

Lolive, J. and Soubeyran, O. (dir.) (2007) *L'Emergence des cosmopolitiques*, Paris: Editions La Découverte.

—— and Tricot A. (2002) 'La constitution d'un réseau d'expertise environne-mentale', *Metropolis*, 108/109: 62–9.

Lu, H.-Y. C. and Morrell, P. (2001) 'Evaluation and implications of environmental charges on commercial flights', *Transport Reviews*, 21(3): 377–95.

Martinez, M. (2001) 'Le prix du bruit autour de Roissy', *Etudes Foncières*, 90: 21–3.

Morrell, P. and Lu, H.-Y. C. (2000), 'Aircraft noise social cost and charge mechanisms – a case study of Amsterdam Airport Schiphol', *Transportation Research Part D*, 5:305–20.

Navrud, S. (2002) *The State-Of-The-Art on Economic Valuation of Noise*, Department of Economics and Social Sciences, Agricultural University of Norway, Report to the European Commission DG Environment.

Offner, J. M. (1993) 'Les effets structurants du transport: mythe politique, mystification scientifique', in *L'Espace géographique*, Doin edn, no.3: 233–42.

Pearce, D. W. and Pearce, B. (2000) *Setting Environmental Taxes For Aircraft: A Case Study of the UK*, Centre for Social and Economic Research on the Global Environment (CSERGE), University London College, Working Paper 2000–26.

Periáñez, M. (2001) *Analyse secondaire de 84 entretiens qualitatifs issus de trois pré-enquêtes psychosociologiques de 1998 portant sur le vécu des situations sonores par les riverains des aéroports d'Orly et de Roissy-CdG*. Report of IPSHA to French Environmental Agency.

Sack, D. (2001) 'Glokalisierung, politische Beteiligung und Protestmobilisierung. Zum Mediationsverfahren Flughafenerweiterung Frankfurt am Main', in *Globalisierung – Partizipation – Protest*, Oplade.

Schipper, Y. (1997) 'On the Valuation of Aircraft Noise-a Meta Analysis', Tinbergen Institute, *PhD Research Bulletin*, 9(2): 1–18.

—— Rietveld, P. and Nijkamp, P. (2001) 'Environmental externalities in air transport markets', *Journal of Air Transport Management*, 7: 169–79.

Stengers, I. (1997) *Sciences et pouvoirs. La démocratie face à la technoscience*, Paris: La Découverte, Coll. Sciences Sociétés.

Thévenot, L. (2006) *L'Action au pluriel: sociologie des régimes d'engagement*, Paris: La Découverte.

Urry, J. (2000) *Sociology beyond Societies. Mobilities of the Twenty-First Century*, London: Routledge.

—— (2003) *Global Complexity*, Cambridge: Polity Press.

Vallet, M., Vincent, B. and Olivier, D. (2000) *La gêne due au bruit des avions autour des aéroports, vol.1 Analyse de la gene*. Report LTE 9920 (French Institute on Transport) to Ministry of Ecology.

Van Praag, B. M. S. and Baarsma, B. E. (2000) 'The shadow price of aircraft noise nuisance: a new approach to the internalization of externalities', Tinbergen Institute Discussion Paper, TI 2001–010/3.

World Health Organization – Pan European Programme (2004) *Transport-related Health Effects with a Particular Focus on Children. Toward an Integrated Assessment of Their Costs and Benefits*. WHO Report.

12 Helipads, heliports and urban air space

Governing the contested infrastructure of helicopter travel

Saulo Cwerner

Introduction

This chapter deals with the effects of urban helicopter traffic, with particular emphasis on the politics, governance and infrastructure of urban flight. Through an analysis of the urban embodiment of helicopter travel in one of the world's greatest metropolises, São Paulo,[1] the chapter shows how new aeromobilities become contested and politicized, and therefore an important feature of the social divisions that mark contemporary mobility. At first sight, as a subject matter for aeromobility research, helicopter flights belong, seemingly, to a world quite apart from the ones analyzed elsewhere in this volume. However, as this chapter shows, many of the aeromobility issues discussed here share a common ground with many of the other chapters. In any case, it is important to start by understanding the specificity of helicopter flights in the context of the study of aeromobilities in general.

Social research into aeromobilities has been fuelled chiefly by the historical impact of the branch of commercial aviation symbolized by the world's great air carriers and their scheduled flights. This strong emphasis on the infrastructure and the social effects of scheduled airline flights is no doubt justified: the rise of mass aviation is a major factor in the globalization of social life, and the great international airports and the hub-and-spoke networks used by major airlines have come to embody the mobile patterns of the contemporary world (see Urry (Chapter 1); Kesselring (Chapter 2); and Derudder *et al.* (Chapter 4), in this volume). Mass aviation and its scheduled flights are certainly the most visible chunk of civil aviation. This is the world of aeromobility that most of us are familiar with, epitomized by busy international airports such as Heathrow, Frankfurt and Hong Kong; and by large, wide-bodied, long-range aircraft such as the Boeing 747 and the Airbus A340.

However, the world of aeromobility is much more varied than the images suggested above might represent. In fact scheduled flights constitute perhaps a minority of the world's air traffic. Distinct from this is the other, larger category of civil aviation, *general aviation*, which includes a gamut of activities from paragliding, ballooning and parachuting to flight training, air ambulance, leisure flights and air cargo. General aviation also includes fixed wing aircraft used for personal or business transportation, as well as the vast majority of helicopter operations. It is clear from this list that there is much more to aeromobilities than ordinarily meets the eye.

As a whole, general aviation caters for more specialized, individualized and ad hoc aeromobility needs, from tackling forest fires to transporting a CEO to an important meeting on the company's own business jet. Conceptually, therefore, it is very hard to determine beyond its variegated form. It is not my intention here to deal with general aviation as a whole, but to consider, through an analysis of the infrastructure of urban helicopter travel, some aspects that give it a more pervasive role in contemporary society.

In order to do that, one needs to focus on one segment of general aviation commonly referred to as *business aviation*. Whether in its commercial ('air taxi') or corporate functions, business aviation provides a distinct, but complementary, network that is even larger in size, and more diverse, than the hub-and-spoke network of scheduled airlines, bringing thousands of smaller airports and locations into the system of aeromobility that enables increasingly globalized economic and social relations. Globally, this sector, comprising mainly what is usually referred to as 'business aircraft' or 'executive jets', is experiencing massive growth, with deliveries in 2007 expected to exceed 1,000 for the first time in the industry's history (Honeywell Aerospace 2006). In Europe, in 2005, business aviation accounted for nearly 7 per cent of all jet flights, having grown twice as fast as the rest of the traffic since 2001 on the back of an imported (from the USA) business model driven by the global economy, increasing profits and prosperity, and a more general acceptance of its economic advantages (Marsh 2006: 4).

The global helicopter industry (in its civil uses) is also in part dependent on business aviation, with corporate applications, alongside emergency and medical services and law enforcement, accounting for over 60 per cent of projected sales in the industry over the next few years (Honeywell Aerospace 2007). Business aviation, either fixed wing aircraft or helicopters, caters for a great number of perceived corporate needs. In 1999 the National Business Aviation Association (NBAA) in the USA identified at least eighteen utilization strategies for business aircraft, following research conducted that year with corporate decision-makers, from key employee travel and customer visits, to making airline connections and attracting and retaining key people (Almy 1999). Business aviation provides strategic network capabilities to an increasing number of corporations and industries, playing a growing role in shaping, through aeromobility, the social relations that underpin the global economy. The helicopter is also analyzed on the basis of the NBAA research results and, not surprisingly, its door-to-door travel capabilities are emphasized: 'The shortest distance between two points (and the most efficient way to travel) is a straight line, which usually can be traveled only by flying via helicopter' (Almy 1999: 13).

This door-to-door (or point-to-point) capability of the helicopter, when used for the purposes of travel across or within urban areas, has had a great impact on the fabric of the city. Where providing business and personal transportation into and out of urban environments, helicopters have necessitated a distinctive infrastructure that is different from that of fixed wing aircraft. This chapter will deal with the politics of such an infrastructure, comprising the evolving political, regulatory and governance structures and systems that enable helicopter traffic over densely populated urban areas. The empirical focus will be the city where this particular form of aeromobility has seen an unprecedented expansion in the past fifteen years: São Paulo, Brazil. The next section examines São Paulo's rise to the condition of the world's helicopter capital. I will then examine the nature of the infrastructure that has enabled such a rise, with a particular focus on the urban network of heliports and helipads. A further section will analyze the politics of urban air travel and the diverse claims over urban airspace, before I conclude with an examination of the lessons that can be drawn from the São Paulo helicopter experiment for aeromobilities theory and research.

The rise of urban helicopter travel in São Paulo

As with business (and general) aviation more broadly, helicopter operations cater for the aeromobility needs and interests of a more specialized, and often socially privileged, group of people. Together with its more common public uses in law enforcement, rescue and medical services (air ambulances), and the increasingly ubiquitous news coverage from the air, helicopters have been particularly suited for transporting workers and visitors to hard to reach locations such as offshore rigs. Since the 1960s, after the introduction of the turbine engine, helicopters have found an important niche in business aviation (Carey 1986), enabling door-to-door aeromobility over short to medium distances. More specifically, very soon in the modern helicopter's history, its vocation as an urban mode of transport was recognized and tested.

I have shown elsewhere (Cwerner 2006) that the urban capabilities of helicopters cannot be dissociated from their technological history and from their place in the wider cultural history of powered flight, which has for more than a century now been searching for personalized modes of air transportation. Helicopters, as the most successful vertical take-off and landing (VTOL) aircraft to date, require much less space for their operations than fixed wing aircraft, and the image of ordinary families using their back gardens as bases for personal helicopter operations was already in the minds of helicopter pioneers in the 1940s (Morris 1945). Nevertheless, a number of factors have contributed to helicopters not enjoying the mass usage that many of the early developers and commentators envisaged (Taylor 1995), among them high purchase and operational costs and perceived safety issues. Overall, the size of helicopter operations has fallen well short of the potential they have been associated with, particularly regarding increases in business efficiency that an alternative to current delays in both surface and air transport could offer (Royal Aeronautical Society 2002: 6).

In light of the factors that have limited the spread of helicopter operations, the experience of one particular global metropolis is instructive. São Paulo has seen a faster expansion in urban helicopter operations than anywhere else in the world in the past fifteen years. Although a great number of these operations are related to the public services, the majority of the urban helicopter traffic there is due to a large increase in business aviation uses, both corporate and commercial. The use of helicopters by public services is certainly an area that merits further research, especially in their media and law enforcement guises, because of the social, cultural and political effects of surveillance, police operations and news reporting from the air. However, from the point of view of its impact on the built fabric and the residents of the city, it is the use of helicopters in business aviation that will be the focus of this chapter. While law enforcement and news reporting make greater use of the helicopter hover capability, business aviation uses its potential for point-to-point transportation more fully. It is particularly this capability that has transformed the airspaces in São Paulo, creating new systems of airspace control and navigation as well as a very distinctive network of heliports and rooftop helipads.

What are the factors that transform the helicopter's aeromobility into a feasible alternative for transportation into, out of, and around cities? The best way of beginning to answer this crucial question is to examine the demand side. Surely helicopter operations are far from constituting a mass mode of air travel. The purchase, ownership (even fractional ownership) and charter of helicopters are beyond the financial means of a very large majority of the population of even the wealthiest nations. For example, the operator US Helicopter, which already operates scheduled helicopter flights between Manhattan heliports and some of New York City's airports for about $170 one-way, introduced a seasonal scheduled helicopter service between New York City and East Hampton Airport in the summer of 2007, serving a number of affluent towns 100 miles from New York. The return weekend trip costs around $1600 for the privilege of a 35-minute direct flight (as opposed to up to three hours by train or even more on congested roads). East Hampton Airport has seen massive growth in operations of both small jets and helicopters in recent years (Belson and Newman 2007). Understandably, beyond the dream of an once-in-a-lifetime panoramic helicopter leisure flight, most people do not entertain the proposition of regular helicopter travel.

However, for an increasing number of people it is just that. Although one does not find references to helicopter charter in the average guidebook, *Worth*, 'the magazine that examines the concerns of families with substantial wealth', in its March 2006 edition featured a large section about executive travel to and in São Paulo. Among extensive information about security, personal safety and other travel issues, the magazine carried an article on 'Private Aviation', including information about international business aviation flights to Brazil and a section on helicopter travel in São Paulo, noting it as a 'popular alternative' to transportation from airport to hotel among top wealthy travellers. Despite the socially limited nature of urban helicopter transportation, it appeared to figure as popular among the wealthiest segment of air travellers. Contributing decisively to that, as this article in *Worth* points out, is the fact that 'São Paulo features nearly 300 heliports in and around the city (compared with about 60 in New York)', going on to report that in São Paulo helicopter traffic is more congested than in any city in the world (Seaton 2006).[2]

The hype about helicopter travel in São Paulo has been going on for a while, at least since the late 1990s, when it became clear that something rather unprecedented was already developing above the city's built fabric, and a spate of local press articles was later reflected in the international coverage of the phenomenon (see, for instance, Bellos (2000), Romero (2000) and Faiola (2002); and Cwerner (2006) for a fuller account of the expansion of São Paulo's helicopter traffic). Urban helicopter travel (comprising operations that either originate or end in an urban area, or both) responds to particular needs (or perceived needs) associated with economic savings (at a level where business aviation makes economic sense, of course), speed of transportation, private safety and exclusive lifestyles. São Paulo's regional economic prominence as the South American global city *par excellence*, together with its social problems and relative lack of investment in transport infrastructure (see Jacobi 2007, Schiffer 2002), have provided the social and economic background for the rise of urban aeromobility.

In the second half of the 1990s, the Brazilian fleet of civilian helicopters experienced an average yearly growth rate of over 11 per cent. Although that astonishing expansion was not matched in the following years, in 2007 the

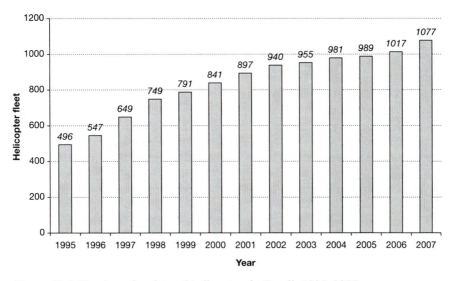

Figure 12.1 Number of registered helicopters in Brazil, 1995–2007.

Source: ANAC (Agência Nacional de Aviação Civil).

figures pointed towards a new period of high growth, in line with global forecasts, with a 6 per cent increase over the previous year in the first nine months alone (Figure 12.1). The growth in the Brazilian fleet since the mid 1990s has been astonishing. Between 1995 and 2006 it grew around 105 per cent, compared with 57 per cent growth in the USA fleet of civilian helicopters in the same period.[3] Around 46 per cent of the fleet are registered in São Paulo State, but it is believed that as many as 400 of the state's 467 helicopters operate in São Paulo's metropolitan area, making it one of the world's three largest urban fleets, alongside New York City and Tokyo.

The nature of helicopter operations means that it is very difficult to obtain precise statistics about their overall volume and direction (that is, departure and destination points as well as other journey details). First, most journeys occur according to visual flight rules (VFRs), that is, the set of aviation regulations that allow aircraft to be operated under weather conditions that provide pilots with safe navigation by visual reference to landmarks and other physical characteristics of the environment. Unless helicopters are operating in more tightly controlled airspace, such as in or around airports, they do not generally fall under strict air traffic control (ATC) rules, although wider regulations do apply (e.g. minimum visibility and altitude parameters). Since a great deal of helicopter operations worldwide do not require clearance from ATC authorities, official statistics about these operations are rare and usually restricted to publicly owned heliports.

Second, a great number of helicopter operations make use of private landing sites. These can be as ordinary places as the gardens of private houses and commercial premises. There are no official statistics related to these operations practically anywhere in the world where there are a considerable number of helicopter operations. Beyond their use by public services and as the mainstay of public transport in the offshore oil and gas industry, helicopters are the spearheads of a very private industry that only recently has been meeting claims for better and more effective monitoring.

For instance, a recent review of helicopter noise in London, UK, commissioned by the Environment Committee of the London Assembly, noted that data collection and monitoring of helicopter movements are very patchy and usually spread over a number of organizations that seem to disagree even over 'which types of helicopter operations are responsible for increases in activity' (GLA 2006: 8). Even where detailed data seem to exist regarding helicopter operations (e.g. regarding the total number of hours flown) as is the case with the USA (FAA n.d.), such data are actually based on sample surveys and estimations: 'with small groups like rotorcraft the estimates are heavily influenced not only by the number of respondents but also by who responds' (FAA 2004: VI-2). Similar difficulties are encountered when attempting to assess the impact of urban helicopter flights in a city such as São Paulo, problems compounded by the fact that many of the city's private landing sites alluded to above are in fact helipads situated on the top of corporate office high-rise buildings, in what is the world's largest urban concentration of rooftop helipads.

Despite the methodological difficulties in trying to establish a clear picture of helicopter operations in the city (see, for more details, Cwerner (2006: 206–7)), it is relatively safe to say that São Paulo has the world's densest urban helicopter traffic. The large concentration of wealth in the city (and some of the surrounding neighbourhoods in the metropolitan area), where the Brazilian (and sometimes South American) headquarters of several national and multi-national corporations are located (Schiffer 2002), together with congested roads and the increasing fear of street violence, have certainly contributed significantly to the rise in demand for this private mode of urban air transportation. The next section examines the urban context and infrastructure that enable helicopter operations in a city of São Paulo's size.

The infrastructure of urban helicopter operations

As with general aviation in general, the corporate and air taxi uses of helicopters depend on the existence of a demand base that can afford their high purchase and operational costs, and for whom the speed and practicality afforded by this form of aeromobility make economic and/or social sense. In urban contexts, helicopter travel is particularly suited to the sprawling metropolitan areas of global cities, where large financial districts are the destination of a great number of executives and professionals whose global networks recreate on a daily basis the very substance of the world economy.

In these contexts, helicopter travel is seen as an essential aspect of transportation between those districts and the airports that serve the metropolitan area, thus connecting the heart of the city with the aeromobility networks discussed elsewhere in this volume. When air operator US Helicopter recently introduced its scheduled services between Downtown Manhattan Heliport and JFK International Airport, it followed the introduction of facilities in the heliport that allow passengers to check themselves and their bags through to their final destination, 'be it Chicago or Shanghai'. A spokesperson for the Port Authority of New York and New Jersey, the heliport operators, noted that customers would be 'executives in the financial community and downtown who want to save the time it can take to ride to the airport and go through security', as the helicopter drops them precisely at the gate, ready to board their onward flights (McGeehan 2006).

Similarly, it is no surprise that recent debates about heliport provision in London have focused on facilities for East London, thus providing a more direct route into the city's financial districts (GLA 2006). And China, which had taken a negative stance towards general aviation for decades, has recently opened up to the capabilities of helicopters. Amidst increasing cooperation with the West, marked initially by the visit of a business delegation in October 2003 (Biddle 2003), the Chinese government has earmarked Shanghai, the economic powerhouse of the region, for a proof of concept heliport facility, the Shaghai Helicopter Centre, which is meant to spearhead civilian helicopter operations in the country. It is the concentration of financial and economic wealth in global cities and city-regions that provides the basis for the rise of urban helicopter operations as a door-to-door extension to global aeromobility networks. And with its three metropolitan airports and exclusive financial districts, São Paulo has provided the clearest example of such operations to date.

What makes the São Paulo experience unique? Part of the answer lies in the particular economic and social geographies of the city. With a population of over 11 million[4] (out of around 20 million in the Greater São Paulo metropolitan area), São Paulo grew fast and haphazardly over the last 100 years, in part due to successive waves of international and domestic migration, and in part due to ineffective city planning (Rolnik 1999). The result is an urban mosaic where many wealthy residential districts lie adjacent to working-class neighbourhoods, including some of the largest *favelas* in the city. The old city centre also 'decentralized', with financial and other services moving to new corporate districts kilometres away.

There are three major airports within the metropolitan area of Greater São Paulo. Eight kilometres to the south of the São Paulo city centre is the Congonhas airport, which until recently was Brazil's busiest, handling mostly regional and national traffic, while 6 kilometres to the north of the city centre is the Campo de Marte airport, used mainly for general aviation purposes, including small business jets and helicopters. Both these airports lie within the city's boundaries, surrounded by densely populated districts. The metropolitan area's international airport, Guarulhos, is located north-east of the city, in the municipality of Guarulhos, around 25 kilometres from the city centre. Most corporate districts are situated between the Congonhas airport and the city centre, notably on and around the Paulista, Faria Lima and Berrini avenues. These districts are also adjacent to some of the wealthiest residential areas in São Paulo (see Figure 12.2). Also of some importance, as far as urban helicopter travel is concerned, is the fact that São Paulo is only some 60 kilometres from the sea. It is not surprising that many wealthy families use helicopters as means of personal transportation to the seaside at weekends and holidays.

Therefore, transportation between the metropolitan airports and the financial/corporate districts is one of the most important features of urban aeromobility, together with more traditional commuting from the suburbs, where some of the most notorious of São Paulo gated communities can be found, such as Alphaville, in the municipality of Barueri, in the Greater São Paulo. However, the socio-economic geographies of the city and the needs of the corporate elite cannot in themselves explain the sharp rise in São Paulo's helicopter traffic. One needs to delve into some of the most characteristic features of VTOL aeromobility (in this case, helicopter flights) in order to understand the uniqueness and, as a result, some of the problems and solutions presented by the São Paulo helicopter experience.

Like the fixed wing aircraft that populates most of the social research into aeromobilities, helicopter operations configure particular forms of airspace,

Figure 12.2 Paulista avenue's financial district and Jardins residential
neighbourhood (author's photograph).

which is itself regulated to various degrees and in various forms. Since
helicopter operations usually follow VFRs, helicopters tend to fly at relatively
low altitudes. In and around large global cities, where busy international
airports are usually located, this creates a particular problem, as the airspace
used by helicopters coincides with, or is contiguous to, that used by fixed
wing aircraft. Many large cities, such as London and New York, solve the
problem by creating specific regulations that may include maximum altitudes,
special helicopter routes or 'corridors', aircraft type requirements (e.g. some
cities only allow twin-turbine helicopters) and various degrees of air traffic
control (for an in-depth analysis of airspace control and regulation, see Budd
(Chapter 6), in this volume). Other cities ban non-essential helicopter flights
altogether.

There is also the issue of the impact of helicopter operations in densely
populated areas, especially regarding aircraft noise, which leads sometimes to
regulations being imposed on the way that airspace is used during those
operations (e.g. minimum altitudes and the use of existing freeways and rivers
as references for special helicopter routes). In São Paulo, airspace regulations
evolved in a similar fashion to other major cities. Until 2004, traffic was
coordinated by helicopter pilots themselves, following a number of regulations issued by civil aviation authorities regarding the use of special routes

or corridors. Complaints from residents also led to a small degree of self-regulation by the industry, mostly through a 'pact' issued by the then Association of Helicopter Pilots of São Paulo State (APHESP).[5] With the growth in traffic, and especially with the proliferation of rooftop operations around Congonhas airport, civil aviation authorities decided in 2004 to introduce a controlled area near the airport, in which helicopters would need clearance from a dedicated control tower in the airport in order to operate (see, for more details, Cwerner (2006: 204–5)).

On the other hand, unlike the large airports necessitated by the modern airliner, the ground infrastructure for most helicopter operations requires much less space. Helicopters have the capability to take off and land in almost any terrain and with very little space, provided there is enough clearance for approach. Sites that see more regular usage usually have special landing pads. In Brazil, like in many other countries, civil aviation authorities distinguish between heliports and helipads. While the latter simply provides space for landing and take off, heliports are larger infrastructures equipped with fuelling, maintenance, hangar and other facilities. With the growth in helicopter use, São Paulo saw the construction of several private heliports, all with VIP facilities, to add to the expanding use of the public heliports in the city's main airports, as the Campo de Marte airport grew to become one of the world's largest helicopter bases.

As importantly, the city saw a proliferation of rooftop helipads, dotted on a growing urban landscape of high-rise corporate buildings, office towers and five-star hotels. There is some disagreement between the figures of helipads in São Paulo provided by different organizations. Recent figures provided by the newly reformed civil aviation agency, ANAC, puts the figure at 210 licensed helipads (Credendio e Gallo 2006), although already in 2004, the then association of helicopter pilots of São Paulo State put the figure at 310, with 190 (practically all being rooftop helipads) in the corporate districts around Paulista and Faria Lima avenues alone (Romanelli and Pitta 2004). A committee at São Paulo's city council, recently instituted to make proposals regarding the crisis at Congonhas airport,[6] stated that the number of helipads was 280 in the beginning of 2007. This confusion in part stems from different licensing or regularization processes adopted by the civil aviation and municipal authorities, but also from a good number of clandestine landing pads or those still awaiting licensing, but already operational.

Whichever is the real number, the fact is that the number of rooftop helipads in São Paulo greatly exceeds that of any other city in the world, and that it is this wide availability of landing pads that makes the proposition of urban aeromobility in São Paulo very attractive. Executives and professionals can arrive directly, by air, at their places of work or meetings, radically cutting the time of transportation as well as bypassing the urban chaos below.

São Paulo's network of rooftop helipads, many of which are 'public' helipads used for commercial purposes (such as landing and take-off points for those visiting or working in nearby buildings), constitutes perhaps the

Figure 12.3 Torre 2000 office building in São Paulo (author's photograph).

world's most developed network available for personal, urban, door-to-door air transportation, and represents the closest an urban environment has come to the aeromobility vision of the helicopter pioneers. At the same time, the rooftop helipad celebrates a new form of aeromobility that radicalizes the features of the vertical city in the mould of theories that espouse a three-dimensional urbanism (Yeang 2002).[7] In its relatively short history, the helipad, a platform that creates a clear space for the landing and take-off of helicopters, has evolved from an engineering problem into an issue of concern to architects, especially after these started to integrate them into the top of high-rise buildings (de Voogt 2007). It has now become an aesthetic element in edifices (see Figures 12.3 and 12.4). According to a São Paulo-based helipad designer and engineer, Carlos Freire, helicopters are changing the metropolis and its buildings, and the trend among professionals is to view rooftop helipads as sculptures, and not simply as functional elements of buildings (Finestra 2006).

On the other hand, by making particular city spaces accessible directly from the air, rooftop helipads have helped create an exclusive mobility space for the wealthiest segment of the economic elite in São Paulo. It would be erroneous to analyze the infrastructure of urban aeromobility in São Paulo simply in terms of social exclusion. It is true that helicopter travel perpetuates and, in some respects, symbolizes, the social differences that are inscribed in architecture and urbanism. It deepens and widens the divisive nature of the urban landscape as theorized by Zukin, both as the architecture of social class *and* as the 'entire panorama that we see' (Zukin 1993: 16). One only has to compare the views afforded by Figure 12.2, on one side, and Figures 12.3 and 12.4 on the other, in order to realize the different perspectives that urban aeromobility (and its airspaces) provides to those in the air and those on the ground.

It is also true that helicopters, taken together as tools of transportation, surveillance and reporting, establish new forms of control and partition in the city. As Marcuse pointed out, 'the controlling city tends to be located in (at the top of, physically and symbolically) the high-rise centres of advanced services', including the command centres of large corporations. He goes on to say that,

> those locations, wherever they may be, are crucially tied together by communication and transportation channels which permit an existence insulated from all other parts of the city, if dependent on them. In that sense a 'space of flows', or a 'space of movement', is an apt, if metaphoric, phrase.
>
> (Marcuse 1995)

What the city of helipads and aeromobility does is to give architectural and urbanistic sense and form to that metaphor. Flows now *mean* flows, symbolized by the air flows under the helicopter rotors. Helicopter flights in

São Paulo's corporate districts produce the ultimate space of flows, not as a metaphor, but as real aeromobility.

However, the opening of urban airspace for aeromobility through various degrees and forms of air traffic control, and the provision of access points across the city may also represent a new chapter in the history of aeromobility and, in particular, in the consolidation of helicopters (and other VTOL aircraft) in transportation and other urban uses. In the last few years there have been a series of bills and proposals put forward by city councillors, as well as by the municipality of São Paulo, with a view to regularize the helicopter traffic over the city. One such bill, proposed in 2001 by the then local councillor José Eduardo Cardozo,[8] envisaged a network of emergency helipads throughout the city, including areas of heavy road traffic, public squares and along the city's rivers. In his justification of the bill, councillor Cardozo noted that the lack of helipads in strategic places across the city limited rescue and evacuation services to privileged areas of the city where most helipads are situated.

Therefore, urban aeromobility, as exemplified by the helicopter experience in São Paulo, both reproduces social divisions in the city *and* opens up new discourses about public airspace. The three-dimensional city also invites debates that are little more than a reignition of old dreams associated with helicopter travel and vertical flight more generally, especially the *flying car* (Robb 2005), which is itself part of a long-standing cultural fascination with personal aeromobility.[9] However, before helicopters (and their successors in

Figure 12.4 New Century building in São Paulo (author's photograph).

the world of VTOL) can herald a new era of aeromobility, they must face an increasing politicization of their impact on the city. The next section will examine the emergent politics of helicopter flights, as well as the changing forms of governance of urban airspace.

The politics and governance of helicopter flights

Owing to their relative obscurity when compared with the fixed wing aircraft operations of major airlines, helicopter flights would not be expected to invite a great deal of political discussion. It would seem that, once the technical and infrastructure issues are resolved, helicopters should enjoy a free ride. However, especially in the case of their urban operations, helicopters have become the subject of much debate and campaigning both at government level and in grass-roots movements.

One of the major issues that affect helicopter operations worldwide is the lack of public acceptance and recognition. Apart from public service operations, such as in rescue and medical evacuation, and specialist transport uses, such as in the offshore oil and gas industry, helicopters tend to draw, at best, indifference or, at worst, outright hostility among regulators and politicians. The difficulties facing the helicopter industry in its attempt to expand helicopter use, especially in corporate and commercial operations, is exemplified very well by developments in European airspace. It is suggested that 'Europe's industry remains a long way off achieving a view among air traffic services and regulators that integrates rotorcraft into civil air operations in a coordinated way and fully utilizes the unique capabilities of vertical lift' (McKenna 2007: 28). And convincing these authorities of the importance of integrating helicopter operations into airspace regulations more fully will be nigh impossible 'as long as politicians are attacking helicopters and publicly opposing expansion of their use (as they have over the last year in London and Paris)' (McKenna 2007: 31).

The question then becomes: why are politicians 'attacking helicopters and publicly opposing expansion of their use', not only in London and Paris, but in many other cities and localities around the world? What are the issues that make helicopters such a *politicized* subject? And what kind of politics develops around urban helicopter operations? In this chapter I will discuss two aspects of helicopter flights that have made them, at the very least, a controversial tool for aeromobility. Although these issues pertain to all helicopters, they tend to become politicized only regarding corporate and commercial helicopter operations, including sightseeing flights, since there is a large degree of public tolerance towards the urban use of helicopters for law enforcement, rescue and medical evacuation. The first issue regards perceptions of public safety when helicopters fly over densely populated areas. The second, and more intricate, issue regards the impact of helicopter noise.

Both concerns over public safety and aircraft noise can be considered as environmental issues in a wider sense, and thus carry the dilemmas and difficulties of articulation that characterize much of environmental politics, especially the contrast between conservationism and policies devoted to effect attitudinal and behavioural change within a liberal-capitalist framework. In the particular case of helicopters' environmental impact, this dualism is expressed by the distinction between demands for bans on flights and policies that favour incentives for the use of newer, quieter aircraft (such as NOTAR or Fenestron-equipped helicopters, which use tail rotor technologies that produce less noise and are deemed safer). Much of the politics around helicopter noise also suffers from the difficulty of articulating demands in the context of wider environmental needs, with charges of NIMBYism frequently cast at campaigning groups. This aspect of environmental politics has been present particularly strongly in the case of groups campaigning against airport expansion.[10]

Issues of public safety abound in discourses about helicopter operations in densely populated areas. In great part this follows a deeply embedded mistrust of such aircraft, which has led to periodic examinations of the safety records of helicopter operations. In fact, the evidence from around the world shows that such safety record varies considerably, but also that public perceptions tend to exaggerate reality. A recent study of the safety record of UK offshore operations revealed it to be on a par with that of car use, although it compares badly with rail and air travel more generally (John Burt Associates Limited 2007: 5). In Brazil, the doubling of the number of helicopter registrations and massive growth in the number of operations in the past decade were

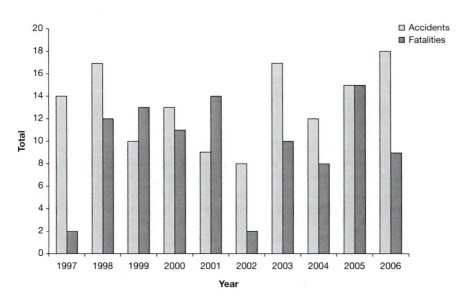

Figure 12.5 Accidents and fatalities in helicopter operations in Brazil, 1997–2006.
Source: CENIPA.

accompanied by an oscillating, but relatively stable, level of both accidents and fatalities, as Figure 12.5 shows. Data from Brazil also reveal that 90 per cent of all helicopter accidents involve human factors, at either operational or maintenance levels, and in more than half of accidents the pilot's judgement is a contributing factor (Tavares 2007). As a result of that, a great deal of regulatory and other initiatives in civil aviation regarding helicopters, both in Brazil and in many other countries, aim at increasing safety awareness and training among pilots.

What then explains the gap between perceptions and reality regarding the safety of helicopter operations? Part of it could be explained by resorting to the psychology of aviation, which is well beyond the scope of this chapter. One simpler explanation derives from the high profile of helicopter accidents. Although the chances of survival here are greater than with accidents involving fixed wing aircraft, helicopter accidents tend to affect proportionally more public personalities because of the exclusive nature of commercial, corporate and some other operations: politicians, business people and other celebrities. Although comparatively rare, helicopter accidents tend to greatly increase and reinforce ordinary people's perceptions of such machines. As with aviation in general, concerns over safety become exposed whenever accidents are reported. For example, on 1 November 2007, no less than three helicopter crashes occurred in the state of São Paulo, two of them in the metropolitan area of Greater São Paulo, one of these leading to three fatalities. The extensive news reporting that ensued was populated by accounts of the pilot in the latter avoiding a busy road when attempting an emergency landing, and of one passenger actually jumping off the aircraft before it crashed into the ground (Soares *et al.* 2007). The accident occurred next to one of the metropolitan area's major heliports, Helipark, from where the helicopter took off. Similar concerns have been raised after a number of crashes involving sightseeing tours in the USA (Klein 2006). The dramatic nature of helicopter accidents, coupled with the high profile of many victims, heightens general perceptions of the public safety of helicopter operations.[11]

The issue of helicopter noise is a more complex one, and has been the subject of numerous reports and campaigns over the years, many of them prompted by state authorities. Claims and concerns over the noise impact of helicopters are usually voiced by local residents and resident associations. In the USA, the New York City Helicopter Noise Coalition has been variably successful in campaigning for restrictions to sightseeing flights over the city, and its activities mirror other groups in the country. In São Paulo, the fight against helicopter noise is led by resident associations of neighbourhoods adjacent to the corporate and financial districts, which experience a great deal of helicopter operations in the city, especially in the Jardins area, a large, middle- to upper-class neighbourhood that extends westwards from Paulista avenue, where there is a large concentration of high-rise buildings, through a lush core of luxurious, low-lying residential homes that end in the Faria Lima avenue, and beyond, to the Pinheiros river, itself one of the city's busiest helicopter 'corridors'.

These resident associations have used a diversity of tactics in São Paulo, including raising awareness locally about noise and other urban aeromobility issues; lobbying members of local government as well as civil aviation authorities; producing independent knowledge through research, e.g. taking noise impact measures; promoting consultations of local residents; initiating lawsuits; and responding to public consultations regarding new legislation. The industry itself has reacted in various ways to such grass-roots movements, often by promoting self-regulation. The Helicopter Association International's (HAI) *Fly Neighbourly* programmes (adopted in São Paulo by the ABRAPHE, the association of helicopter pilots) are examples of such strategies, through which pilots and operators alike are encouraged to adopt noise abatement measures (moves that are often greeted with suspicion by grass-roots movements[12]).

One of the main problems that the grass-roots environmental and residential groups have encountered has been the striking lack of instituted mechanisms of representation in the local governance of aviation and airspace issues (related to both helicopter and fixed wing aircraft, as well as to the infrastructure of airports, heliports and helipads). Traditionally, aviation matters have been the sole concern of civil aviation authorities.[13] As a result of the expansion of aviation and its local impact on cities and regions, there has been a worldwide shift towards finding new forms of airspace governance. This shift has been particularly strong regarding helicopter issues, because of their closer and more direct impact on the urban environment.

In São Paulo, like in many other cities (see GLA 2006), the local authority has finally caught up with the three-dimensional city and has been attempting to expand its jurisdiction over urban airspace. Several bills introduced by successive administrations, as well as by a number of local councillors, have attempted to regulate aspects of helicopter flights over the city, including the regularization of helipads and heliports, and restrictions over flight hours and zones. This process has seen the superimposition of discourses about resident welfare (noise pollution) over traditional concerns of civil aviation, such as air traffic control and flight safety, in what I call the *municipalization* of airspace. A good example of that is given by attempts by the São Paulo municipality to legislate over the licensing of helipads. New regulations impose conditions that were absent from licensing procedures dictated by civil aviation rules, such as the production of neighbourhood and local environmental impact reports.

Together with new proposed airspace legislation, local authorities, often spurred by grass-roots movements, are seeking to institute new governance structures, to increase their say in local aviation matters. However, this process, as argued in relation to the expansion of aviation in the UK (Griggs and Howarth 2006), is more complex than naïve views of 'governance' may suggest, because it is clear that the civil aviation authorities, together with the industry, still possess a great amount of power in determining the outcome of consultations and policy-making.

Conclusion: the politics of aeromobility

One of the paradoxes of the contemporary environmental politics of aviation is the fact that, although people are increasingly concerned about the environmental impact of aviation, demand for air travel is rising sharply and is forecast to continue to do so. How do we explain this? Can analyses of the growth and the politics of urban helicopter flights help us answer this question and thus indicate something more general about contemporary aeromobility? Or are helicopters, and their relatives in general aviation, little more than a fancy, though prestigious, excrescence to the world of aviation?

There is no doubt that, somehow shielded and relatively hidden away from mass aviation and the airspaces it produces, general aviation, and business aviation in particular, plays a fundamental role in contemporary society, enabling the globalizing networks that run and articulate in many respects the global economy. In a world where speed has become perhaps the ultimate value (Chesneaux 2000), personal aeromobility is a luxury that makes economic as well as social sense. However, the social exclusiveness of helicopter travel mirrors that of the first few decades of powered flight. Mass aviation is a relatively recent phenomenon. Until the 1950s, aeromobility was indeed a preserve of the few (see Hudson 1972). It is difficult to predict the long-term development of vertical flight, let alone of aviation in general, so it is not wise

Figure 12.6 Aeromobility and the view from above (author's photograph).

to speculate whether the massification of personal aeromobility is a chimera that belongs to the past rather than a genuine prospect. It is also important to bear in mind that the fascination with personal and more exclusive forms of aeromobility far exceeds the population segment that it regularly serves, which is attested by the growing popularity of helicopter sightseeing flights. Vertical flight affords a view of nature and the lived environment of cities that no other form of aeromobility can (see Figure 12.6).

At the same time, the environmental impact of the demand and desire for personal aeromobility far outweighs the direct benefit it produces to its users. This discrepancy between the number of people adversely affected and those who benefit directly from such mobility is much higher than the one that can be observed in mass aviation. In the case of urban helicopter flights, the disproportionate nature of such an impact is even more evident, especially regarding aircraft noise. Paradoxically, it is this form of aeromobility that becomes the subject of a local politics in which institutions and groups totally unrelated to the industry become stakeholders in claims for jurisdiction over the airspaces produced by this aeromobility. That does not mean that local groups and governments have little stake in matters related to mass aviation. However negative, the impacts of large airports are spread in varied forms over large territories and are often offset by perceived gains and benefits locally, while the airspaces of urban helicopter flight are much closer to the mobilized neighbourhoods. In this sense, the politics of urban aeromobility provides a test tube for wider processes of the reclamation of airspaces as objects of democratic politics, citizenship and participation. It constitutes a model for the politicization of airspace, where different claims and issues that were traditionally treated separately come together in an evolving governance of urban airspace.

São Paulo is perhaps the first metropolis that has, in one way or another, adopted the dream of the *flying car* through the production of a unique urban airspace. At the same time, it has generated a great deal of debate over how such airspace is (to be) regulated and legislated upon. At a time when the social sciences and humanities have finally caught up with the need to know more about airspaces, the São Paulo experience confirms that research into aeromobility inevitably transcends questions of what we know, to embrace the increasingly fundamental problem of *what we do about it*.

Notes

1 The bulk of the research carried out for this chapter was conducted from April to September 2004 in São Paulo, Brazil, and was generously funded by the United Kingdom's Economic and Social Research Council (ESRC Award Number RES-000–22–0732). The project involved collection of secondary data, including media reports, over fifty semi-structured interviews with a diverse set of stakeholders, and a representative survey of the population of some of the city's districts most affected by helicopter traffic. The author would like to thank the ESRC for making this research possible.

2 It is important to note that this article uses the word 'heliports' to refer to what are, in most cases, helipads. The distinction will be made clear in the next section.
3 Source: Federal Aviation Administration (see FAA n.d.: I-4).
4 Source: IBGE (estimate for July 2006).
5 APHESP became the Brazilian association for Helicopter Pilots (ABRAPHE) in 2005.
6 The crisis at Congonhas followed the crash of TAM flight 3054 on 17 July 2007, Latin America's biggest ever air crash, with the loss of 199 lives, among crew, passengers and people on the ground. Air safety and civil aviation management have been under scrutiny in Brazil since a mid-air collision between a passenger airliner and an executive jet over the Amazon in September 2006, which resulted in the death of all aboard the airliner.
7 The vertical city has figured prominently in the cultural history of aeromobilities, from airships docking on the Empire State Building in New York City to an extensive imagination in both literature and architecture. See, for instance, Pascoe (2003: 68–70).
8 Bill 01–0554/2001 in the São Paulo City Council (Câmara Municipal de São Paulo).
9 For an analysis of the collapse of the dream of personal flying in post-war USA; see Trimble (1995).
10 But see Griggs and Howarth (2004) for a more successful case of universalizing local campaigns.
11 This argument reflects the views of helicopter pilots and operators interviewed in 2004.
12 See Moorman (2005).
13 In Brazil this democratic deficit is compounded by the fact that civil aviation management and regulation are still overseen by the military.

Bibliography

Almy, D. W. (1999) *Business Aircraft Utilization Strategies*. National Business Aviation Association.
Bellos, A. (2000) 'Rich Brazilians look down on crime and traffic', *The Guardian*, 7 August 2000.
Belson, K. and Newman, M. (2007) 'Schumer calls for limits on helicopters to Hamptons', *The New York Times*, 10 July. Available at www.nytimes.com/2007/07/10/nyregion/10heli.html?n=Top/Reference/Times%20Topics/Organizations/F/Federal%20Aviation%20Administration (accessed 16 October 2007).
Biddle, T. M. (2003) 'Where are the helicopters? A report on the First Shanghai Delta Helicopter Business Delegation October 12–15, 2003', *Rotor* Winter 2003–2004: 8–9.
Carey, K. (1986) *The Helicopter: an Illustrated History*, Wellingborough: Patrick Stevens.
Chesneaux, J. (2000) 'Speed and democracy: an uneasy dialogue', *Social Science Information/Information sur les Sciences Sociales*, 39(3): 407–20.
Credendio, J. E. and Gallo, R. (2006) 'São Paulo tem 131 helipontos clandestinos', *Folha de São Paulo* 8 May. Available at www1.folha.uol.com.br/folha/cotidiano/ult95u121200.shtml (accessed 16 October 2007).
Cwerner, S. (2006) 'Vertical flight and urban mobilities: the promise and reality of helicopter travel', *Mobilities* 1(2): 191–215.
de Voogt, A. (2007) *Helidrome architecture*, Rotterdam: 010 Publishers.

FAA (n.d.) *General Aviation and Air Taxi Activity and Avionics (GAATAA) Surveys CY2006*, Federal Aviation Administration. Available at www.faa.gov/data_statistics/ aviation_data_statistics/general_aviation/CY2006/media/FINAL%20FAA%202006 %20Chapter%201%20101407.pdf (accessed 16 October 2007).

FAA (2004) *FAA aerospace forecasts: fiscal years 2004–2015*, Federal Aviation Administration. Available at www.faa.gov/data_statistics/aviation/aerospace_forecasts/ 2004-2015/media/FORECAST%20BOOK.pdf (accessed 16 October 2007).

Faiola, A. (2002) 'Brazil's elites fly above their fears', *Washington Post*, 1 June 2002.

Finestra (2006) 'Helipontos: a sua próxima parada', *Finestra* 45. Available at www.arcoweb.com.br/tecnologia/tecnologia71.asp (accessed 16 October 2007).

GLA (2006) *London in a spin – a review of helicopter noise*, London: Greater London Authority.

Griggs, S. and Howarth, D. (2004) 'A transformative political campaign? The new rhetoric of protest against airport expansion in the UK', *Journal of Political Ideologies*, 9(2): 181–201.

—— and —— (2006) 'Airport governance, politics and protest networks', in M. Marcussen and J. Torfing (eds.) *Democratic Network Governance Networks in Europe*, Basingstoke: Palgrave.

Honeywell Aerospace (2007) 'Helicopter five-year purchase forecast', *World Aircraft Sales Magazine*, April: 58–66.

—— and —— (2006) 'Honeywell business aviation outlook 2006–16', *World Aircraft Sales Magazine*, December: 70–90.

Hudson, K. (1972) *Air travel: a social history*, Totowa, New Jersey: Rowman and Littlefield.

Jacobi, P. (2007) 'Two cities in one: diverse images of São Paulo', in K. Segbers (ed.) *The Making of Global City/Regions: Johannesburg, Mumbai/Bombay, São Paulo and Shanghai*, Baltimore: The John Hopkins University Press.

John Burt Associates Limited (2007) *UK Offshore Public Transport Helicopter Safety Record*, Oil and Gas UK.

Klein, D. A. (2006) 'Spate of copter crashes prompts concern', *The New York Times*, 5 February. Available at www.nytimes.com/2006/02/05/travel/05prac.html?_r= 1&oref=slogin (accessed 16 October 2007).

McGeehan, P. (2006) 'New helicopter service promises Wall St. to J.F.K., in 9 minutes', *The New York Times*, 6 February. Available at www.nytimes.com/2006/ 02/06/nyregion/06chopper.html?_r=1&oref=slogin (accessed 16 October 2007).

McKenna, J. T. (2007) 'Pursuing the fair share', *Rotor & Wing*, October 2007: 28–31.

Marcuse. P. (1995) 'Not chaos, but walls: postmodernism and the partitioned city', in S. Watson and K. Gibson (eds.) *Postmodern Cities and Spaces*, Oxford: Blackwell.

Marsh, D. (2006) *Getting to the Point: Business Aviation in Europe*, Brussels: EUROCONTROL.

Moorman, R. W. (2005) 'Good approaches make good neighbours', *Rotor & Wing*, November. Available at www.aviationtoday.com/rw/personalcorporate/exectransport/ 1828.html (accessed 16 October 2007).

Morris, C. L. (1945) *Pioneering the Helicopter*, New York: McGraw-Hill.

Pascoe, D. (2003) *Aircraft*, London: Reaktion Books.

Robb, R. L. (2005) 'Driving on air: 20th century flying carpets', *Vertiflite* 5(1): 2–11.

Rolnik, R. (1999) 'Para além da lei: legislação urbanística e cidadania (São Paulo 1886–1936), in M. A. De Souza, S. C. Lins, M. Do P. C. Santos and M. Da C. Santos (eds.) *Metrópole e Globalização: Conhecendo a Cidade de São Paulo*, São Paulo: Editora Cedesp.

Romanelli, A. and Pitta, I. (2004) 'São Paulo tem 75 per cent dos locais de pouso do país', *O Estado de São Paulo* 19 August: C-3.

Romero, S. (2000) 'Rich Brazilians rise above rush-hour jams', *The New York Times*, 15 February. Available at query.nytimes.com/gst/fullpage.html?res=9C06E4DF 1531F936A25751C0A9669C8B63 (accessed 16 October 2007).

Royal Aeronautical Society (2002) *Expanding the Role of the Helicopter and Related Vertical Flight Aircraft*, London: Royal Aeronautical Society.

Schiffer, S. R. (2002) 'São Paulo: articulating a cross-border region', in S. Sassen (ed.) *Global Networks, Linked Cities*, London: Routledge.

Seaton, M. (2006) 'Private Aviation', *Worth*, March. Online. Available at www. worthmagazine.com/Editorial/Executive-Travel/2006-March/Executive-Travel-Sao-Paulo-Private-Aviation.asp (accessed 16 October 2006).

Soares, A., Henrique, B. and Dacauaziliquá, J. (2007) 'Em 2h, 3 quedas de helicóptero e 3 mortes em SP', *O Estado de São Paulo*, 2 November. Available at www.estado.com. br/editorias/2007/11/02/cid-1.93.3.20071102.1.1.xml (accessed 8 November 2007).

Tavares, B. (2007) 'Frota cresce 66 per cent; acidentes permanecem estáveis', *O Estado de São Paulo*, 2 November. Available at www.estado.com.br/editorias/2007/11/ 02/cid-1.93.3.20071102.2.1.xml (accessed 8 November 2007).

Taylor, J. W. R. (1995) 'Constraints impeding the commercial use of helicopters', in W. M. Leary (ed.) *From Airships to Airbus: the History of Civil and Commercial Aviation (Volume I – Infrastructure and Environment)*, Washington, DC: Smithsonian Institution Press.

Trimble, W. F. (1995) 'The collapse of a dream: lightplane owenership and general aviation in the United States after World War II', in W. F. Trimble (ed.) *From Airships to Airbus: The History of Civil and Commercial Aviation. Volume 2: Pioneers and Operations*, Washington and London: Smithsonian Institution Press.

Yeang, K. (2002) *Reinventing the skyscraper: a vertical theory of urban design*, New York: John Wiley & Sons.

Zukin, S. (1993) *Landscapes of power: from Detroit to Disney World*, Berkeley: University of California Press.

Index

Theories of Race and Racism
A Reader

Les Back, University of London, UK
John Solomos, City University,
London, UK

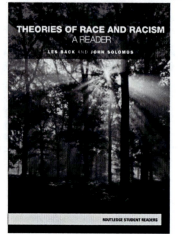

Theories of Race and Racism: A Reader is an important and innovative collection that brings together extracts from the work of scholars, both established and up and coming, who have helped to shape the study of race and racism as an historical and contemporary phenomenon.

This second edition incorporates new contributions and editorial material and allows readers to explore the changing terms of debates about the nature of race and racism in contemporary societies. All six parts are organized around the contributions made by theorists whose work has been influential in shaping theoretical debates. The various contributions have been chosen to reflect different theoretical perspectives and to help readers gain a feel for the changing terms of theoretical debate over time. As well as covering the main concerns of past and recent theoretical debates it provides a glimpse of relatively new areas of interest that are likely to attract more attention in years to come.

Contents
Part 1: Origins and Transformations
Part 2: Sociology, Race and Social Theory
Part 3: Racism and Anti-Semitism
Part 4: Colonialism, Race and the Other
Part 5: Feminism, Difference and Identity
Part 6: Changing Boundaries and Spaces

March 2009
HB: 978-0-415-41253-7: **£85.00**
PB: 978-0-415-41254-4: **£23.99**

Routledge
Taylor & Francis Group

www.routledge.com/sociology

Contemporary Social Theory
An Introduction

Anthony Elliott, Flinders University, Australia

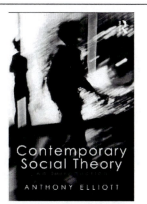

This book is arguably the definitive undergraduate textbook on contemporary social theory. Written by one of the world's most acclaimed social theorists, Anthony Elliott provides a dazzlingly accessible and comprehensive introduction to modern social theory from the Frankfurt School to globalization theories and beyond.

In distilling the essentials of social theory, Elliott reviews the works of major theorists including Theodor Adorno, Herbert Marcuse, Michel Foucault, Jacques Lacan, Jacques Derrida, Anthony Giddens, Pierre Bourdieu, Julia Kristeva, Jurgen Habermas, Judith Butler, Slavoj Zizek, Manuel Castells, Ulrich Beck, Zygmunt Bauman, Giorgio Agamben and Manuel De Landa.

Every social theorist discussed is contextualized in a wider political and historical context, and from which their major contributions to social theory are critically assessed. This book is essential reading for students and professionals in the fields of social theory, sociology and cultural studies, as it is both an original enquiry and a consummate introduction to social theory.

Contents

Routledge
Taylor & Francis Group

September 2008
PB: 978-0-415-38633-3: £24.99
HB: 978-0-415-38632-6: £85.00

www.routledge.com/sociology

Emotions

A Social Science Reader

Monica Greco, Goldsmiths, University of London, UK
Paul Stenner, University of Brighton, UK

Are emotions becoming more conspicuous in contemporary life? Are the social sciences undergoing an an 'affective turn'? This Reader gathers influential and contemporary work in the study of emotion and affective life from across the range of the social sciences. Drawing on both theoretical and empirical research, the collection offers a sense of the diversity of perspectives that have emerged over the last thirty years from a variety of intellectual traditions. Its wide span and trans-disciplinary character is designed to capture the increasing significance of the study of affect and emotion for the social sciences, and to give a sense of how this is played out in the context of specific areas of interest.

The volume is divided into four main parts:
- universals and particulars of affect
- embodying affect
- political economies of affect
- affect, power and justice.

Each main part comprises three sections dedicated to substantive themes, including emotions, history and civilization; emotions and culture; emotions selfhood and identity; emotions and the media; emotions and politics; emotions, space and place, with a final section dedicated to themes of compassion, hate and terror. Each of the twelve sections begins with an editorial introduction that contextualizes the readings and highlights points of comparison across the volume. Cross-national in content, the collection provides an introduction to the key debates, concepts and modes of approach that have been developed by social scientist for the study of emotion and affective life.

Contents
Introduction: Emotion and Social Science **Part 1: Universals and Particulars of Affect.** Emotions, History and Civilization. Emotions and Culture. Emotions and Society **Part 2: Embodying Affect.** Emotions, Selfhood and Identity. Emotions, Space and Place. Emotions and Health **Part 3: Political Economies of Affect.** Emotions in Work and Organizations. Emotions, Economics and Consumer Culture. Emotions and the Media **Part 4: Affect, Power and Justice.** Emotions and Politics. Emotions and Law. Compassion, Hate, and Terror

November 2008
HB: 978-0-415-42563-6: **£95.00**
PB: 978-0-415-42564-3: **£25.99**

www.routledge.com/sociology

Jean Baudrillard

Fatal Theories

David B. Clarke, Marcus Doel, William Merrin, Richard G. Smith all from
The University of Wales, Swansea, UK

Jean Baudrillard was one of the most influential, radical, and visionary thinkers of our age. His ideas have had a profound bearing on countless fields, from art and politics to science and technology. Once hailed as the high priest of postmodernity, Baudrillard's sophisticated theoretical analyses far surpass such simplistic caricatures. Bringing together Baudrillard's most accomplished and perceptive commentators, this book assesses his legacy for the twenty-first century. It includes two outstanding essays by Baudrillard: a remarkable, previously unpublished work entitled 'The vanishing point of communication,' and one of Baudrillard's final texts, 'On disappearance', a veritable tour de force that serves as a culmination of his theoretical trajectory and a provocation to a new generation of thinkers. Employing Baudrillard's key concepts, such as simulation, disappearance, and symbolic exchange, and deploying his most radical strategies, such as escalation, seduction, and fatality, the volume's contributors offer a series of thought-provoking analyses of everything from art to politics, and from laughter to terror. It will be essential reading for anyone concerned with the fate of the world in the new millennium.

Contents

September 2008

 Routledge
Taylor & Francis Group

HB: 978-0-415-46442-0: **£75.00**

www.routledge.com/sociology